UPDATE ON SPACE

VOLUME I

Edited by

B. J. BLUTH
&
S. R. McNEAL

NATIONAL BEHAVIOR SYSTEMS
Granada Hills, California

1981

UPDATE ON SPACE: VOLUME I

National Behavior Systems
11601 Balboa Blvd.
Granada Hills, CA 91344

Library of Congress Catalog Number 80-52460

Bluth, B. J., and S. R. McNeal, Eds
Update on Space

ISBN 0-937654-00-0

10 9 8 7 6 5 4 3 2 1

DEDICATION

This book is dedicated to Earl Blount of Rockwell International whose encouragement and support helped make it possible.

ACKNOWLEDGEMENT

Special thanks go to Dr. Anita Caspary for critically reviewing each of the chapters with painstaking care from the point of view of the non-aerospace reader.

CONTENTS

FOREWORD

We, the people of Earth, have before us the most dynamic challenge of our existence — the cultivation of the space environment. With timely opportunities we can broaden our humanistic growth and universal understanding. We can stimulate economic growth, create new jobs, conceive new careers, develop new products, anticipate increased industry, and imagine new beginnings. We are in embryonic stages of a limitless pattern of growth.

Short term and long term projections point to a New Age, perhaps the most significant stage yet to be taken in human evolution. The next step, however, is not inevitable. We must make the choice to take the risks and build the social and technical vehicles to this new future that includes humanity in space. Such decisions require an informed public provided with reliable and accurate information from those in the best position to know. The *Update on Space* series is intended to provide such access to those hoping to learn and become part of the creation of a human presence in the limitless frontier of space.

The *Update on Space* series is an extension of the Update on Space summer educational program which began in 1978 at California State University, Northridge. The goal of this program was to provide those interested in space with authoritative briefings on the realistic options in space in the near term. Encouraged and supported by people in the aerospace industry, especially Rockwell International, TRW, and Lockheed, speakers with important responsibilities for research and development in space activities were invited to give briefings and meet informally with the program participants. The speakers represented the aerospace industry, NASA, the military, the university community, private research organizations, independent consultants, and other space related enterprises. Participants ranged from high school and college students to stockbrokers, engineers, realtors, teachers from every level, graduate students, and retired people. All the presentations were taped, and the eight synopses that compose the content of this first volume of a series of books were based on those lectures. In the university setting these professionals gave their presentations in language suited to the interested and intelligent lay person, and this conversational tone and general style is purposely preserved in order to open their work to everyone.

Update on Space is the insiders report on our space future. With the builders of Apollo and Skylab, we invite you to look ahead to what man can do in space in our lifetime. What do military people say about war and space and peace? What kinds of materials might soon be made in weightlessness, and how important can they be? Can the high technology of space help the poor people of developing countries or the energy scarce industrial nations? What might life on the Moon really be like? Can people adapt to life in space physiologically and psychologically? These questions are addressed by those currently working in space industries, space research, and space sciences.

We hope these lectures will show you that our real prospects far exceed fantasy. They point to significant accomplishments that could be made in the very near term. Today, we are ready to begin. Please join us on the cutting edge of our future.

B. J. Bluth
and
S. R. McNeal
Northridge, CA

We have to go forward.
We have to improve.
We have to grow and grow up.
Life as we know it is an industrial process.
Man continues this natural evolution.
To reject the technology of life is to reject life itself.

Krafft A. Ehricke

GEORGE V. BUTLER

Director, Advanced Space Concepts,
McDonnell Douglas Astronautics Company

Mr. Butler directs efforts on system studies and new concepts for space vehicles, as well as launch and orbit-to-orbit transportation vehicles. He has directed advanced efforts on Spacelab, Space Station, Space tug, and other manned systems.

Previous hardware assignments included program management roles in the Skylab, Saturn/Apollo and Delta Programs.

Prior to joining McDonnell Douglas, Mr. Butler was Flight Test Manager with Chance Vought on the Regulus missile program.

Mr. Butler is an Aeronautical Engineering graduate. He is an Associate Fellow of the American Institute of Aeronautics and Astronautics (AIAA) and a Senior Member of the American Astronautical Society (AAS).

1

SPACE STATIONS, 1959 — TO?

GEORGE V. BUTLER

The requirement for mankind to build space stations orbiting Earth is tied to two inescapable facts of life. The first is that despite a slowdown in developed nations, the population of the world is still in an explosive state of growth. Second, that man will always demand an increase in his standard of living commensurate with his means and available technology.

Figure 1 illustrates the driving forces of population growth and standard of living, and how these create problems. We can look to space stations for a contribution to the solution of some of these problems.

Space station development has been thought about for a long time, evolving in a series of gradual stages. It is very probable that one day we will have very large space stations where people will live and work in comparable comfort. But it is very unlikely we will start out building habitats for 10,000 people like the ones Professor Gerard O'Neill has suggested. A more realistic scenario for the future would be a step-by-step development that will start with Spacelab stations of 3-4 men. Then

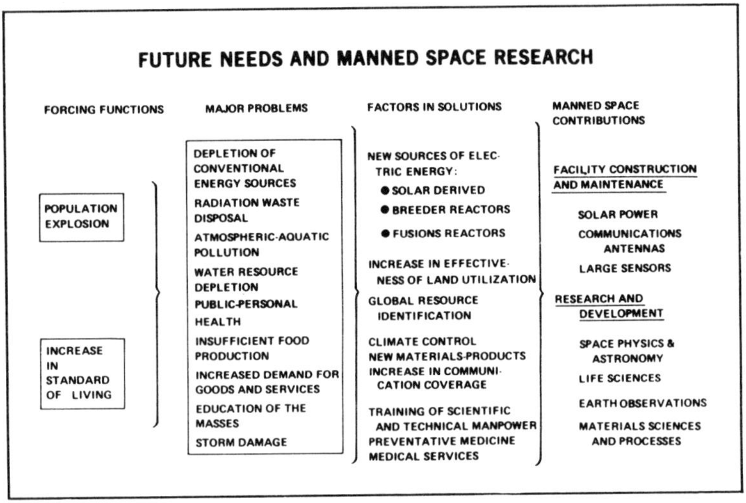

Figure 1

there will gradually emerge expanding crew habitats as space manufacturing and research requirements grow and the pressures of growing population and improving standards of living justify building larger stations.

EARLY SPACE STATION DESIGNS

HERMAN OBERTH'S FIRST SPACE STATION

In 1923, Herman Oberth, a German scientist, wrote what probably was the first technical paper on a space station. This was a manned space station for Earth observations, communications, and a space depot or a stopping-off place for interplanetary flights. He was certainly far ahead of his time since in 1923 there were no large rockets available. Robert Goddard, the U.S. father of rocketry, was just flying his first experimental rocket motors in Massachusetts during the early thirties. It was many years after that before rocketry came anywhere near the point where we could think of putting man out into space.

Figure 2 illustrates that starting in the late fifties, considerable interest began to surface and starting in 1958 through the present time, there have been numerous studies and a few actual hardware manned space habitats.

Figure 3 shows that even the humorists have had their say in the evolutionary process for a space station.

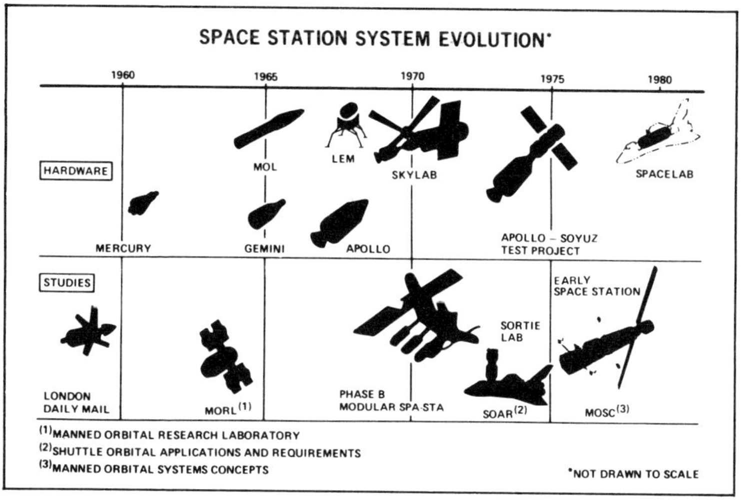

Figure 2

Figure 3

THE LONDON DAILY MAIL PROJECT

The London Daily Mail Study, done in 1958 to 1959, was the first preliminary engineering design for a space station, and it was considered very radical in its time.

The London Daily Mail, a major London newspaper, was sponsoring a "Home Show" and they wanted something very unusual as a central attraction. The theme they selected was "A Home In Space." They held a competition between all the Aerospace Companies (we didn't call them Aerospace Companies then; we called them Aircraft Companies) and Douglas Aircraft Company (now McDonnell Douglas) won that competition for a $10,000 contract. (I'd like to remark that as a company, we have been involved in every major space station study contract that's come along since.) This "Home In Space" idea developed into the concept shown in Figure 4, which is a full size four-man astronomical observatory model with a 60-inch mirror. The fairings on the side were to be used as solar panels when the capsule achieved orbit. The men rode in the forward end of the capsule for take-off and return. While in space, they were to live and work in the tank area which held propellant for launch. After the propellant was expended, and the tank purged, to make it liveable, they were able to climb down into the tank bay.

Douglas built this full-size mockup and shipped it to London where it was on exhibit for two years. It was seventeen feet in diameter and the flight design weighed about 30,000 pounds. It was pretty far ahead of its time because it was many years before the U.S. launched anything even close to 20,000 or 30,000 pounds. The first satellite we launched was under 100 pounds. The London Daily Mail Project, to my knowledge, was the first completely thought-out engineering design for an Earth orbital space station.

Nothing much happened in space for the next few years. The U.S. sent up the first Explorer and we took a couple of shots at the Moon in the early 1960's, but these were all small payloads and had no manned implications. Nobody was thinking seriously about manned payloads that early. But then, in

1961, President Kennedy announced we were going to send men to the Moon which really inspired development of the big rockets and the man-in-orbit thinking that got us into developing space station concepts.

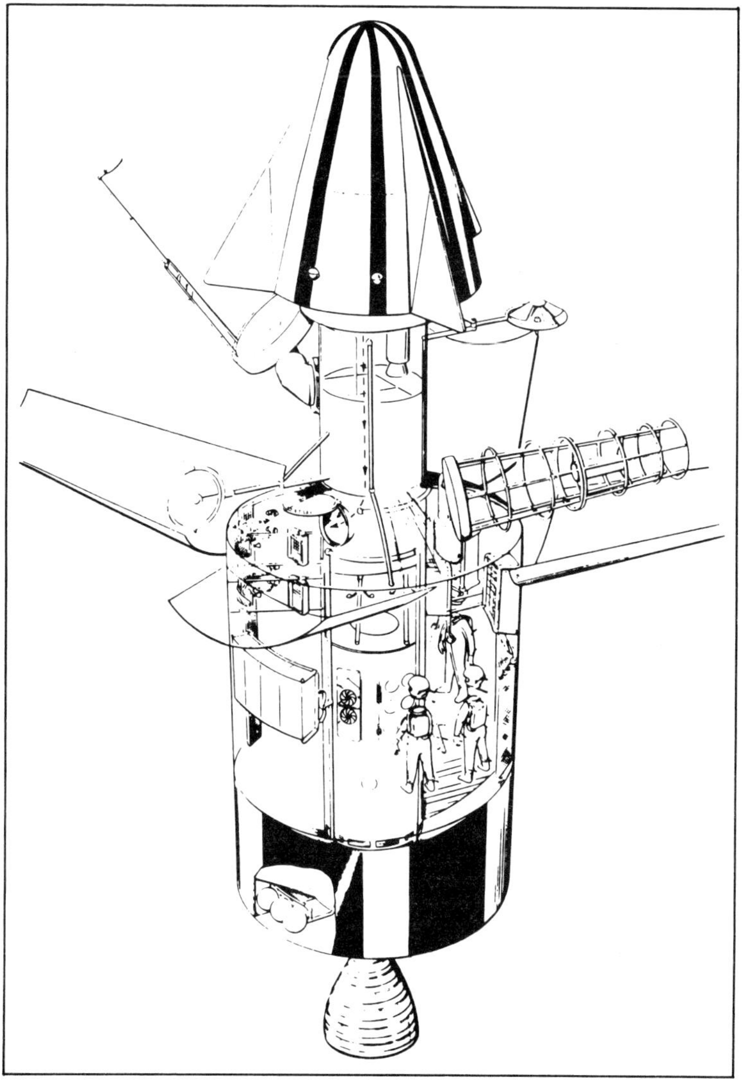

Figure 4

THE MORL

The first major space station design (Figure 5) developed during the Moon mission, was called MORL — Manned Orbiting Research Laboratory. At that point, we were still thinking that the reason one would have a space station was to do fundamental research where the advantages of zero "g" and the available space vacuum would allow experiments to be discovered that could benefit the world and the generations that follow us.

The MORL was an interesting concept because at such an early time, it provided for both zero "g" and one "g" artificial gravity. There was great concern that man could not live for extended periods in zero "g." The final booster stage was to be put into orbit and attached with cables to the station itself. Then they were to be spun together using the booster stage as a counterweight for balance, to provide artificial gravity in the station. The MORL had a 4-6 man crew and it was planned to

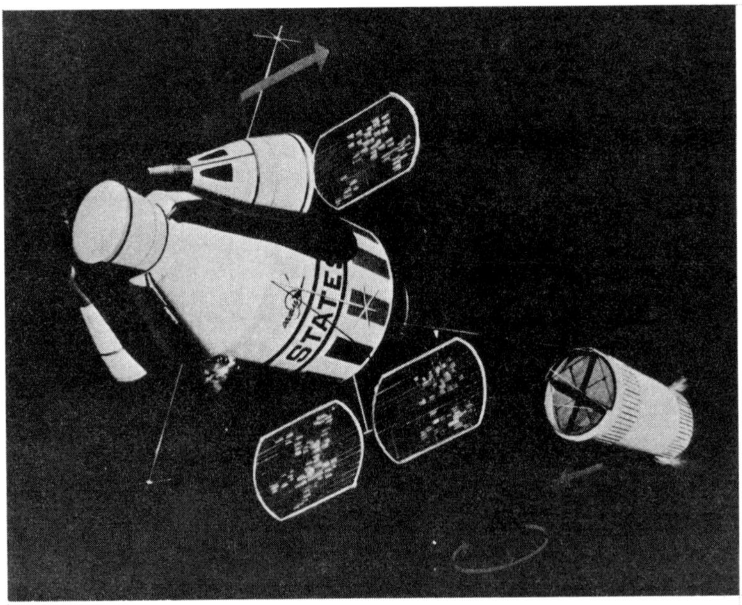

Figure 5

be 22 feet in diameter, which provided quite a large working volume.

THE MOL

The MOL (Manned Orbiting Laboratory) Program, which was a Department of Defense (DOD) Classified Manned Space Program, reached the hardware development stage, but it was cancelled prior to actual flight because of changing requirements.

MAN INTO SPACE

But we did put men in space! You all know about Mercury and Gemini which took our first astronauts around the Earth. If you have ever seen those capsules at close range, you will realize they weren't very comfortable. The Mercury capsule was about the equivalent of putting a cocoon around a motorcycle. It was very confined. Gemini was a very slightly larger size than Mercury, built to accommodate two men. You will recall Frank Borman and James Lovell, who stayed up for two weeks. Apollo, a 3-man capsule, wasn't nearly as cramped, but it was still small. We did not get our first real space station until Skylab.

SKYLAB: OUR FIRST SPACE STATION

The Skylab program started out with many of the same ideas as the 1959 London Daily Mail Astronomical Observatory Project. At that time, we had the first of the Saturn vehicles, (the Saturn 1B) which was used early in the Apollo program for low Earth orbit tests. It was a two-stage launch vehicle. Some of the people that worked on the early Skylab program had also worked on the London Daily Mail Project. Their idea was to take the second stage (the S-IVB) of the early Saturn rocket and use it as a laboratory after the propellant was consumed. This was termed a "wet" workshop. NASA thought that it was a good idea and embarked on a design program to establish the feasibility of the idea.

About three years into the program, the design kept getting more and more complex because people wanted to make a very capable research laboratory out of it, not just a simple workshop. They wanted comfortable living quarters, and room for rather sophisticated experiments, yet, they still wanted to utilize the propellant tank of the launch vehicle used to put the Skylab into orbit. The problems associated with meeting all of these requirements caused the design to become very marginal.

Fortunately, the huge Saturn V rocket was being designed to put Apollo on the Moon and it had sufficient thrust available to put the third stage (the S-IVB) into orbit, now making it possible to design a "dry" Skylab Laboratory. In other words, it was no longer necessary to use the S-IVB as a propulsion stage to get into orbit, making it possible to build a whole sophisticated laboratory, outfitted and checked out, on the ground.

The early 1973 first Skylab launch was nearly a disaster. During the launch into orbit, the protective shield designed to provide meteoroid protection and temperature control, allowed one of the two large solar panels to rip loose from the vehicle, and also jammed the other solar panel. A fairly hectic two weeks followed while engineers tried to figure out how to save the mission. A plan was devised by which Pete Conrad, who had Gemini and Apollo experience, commanded the crew who were sent up with a repair kit to try to salvage the project. A reflective mylar shield with braces that opened out much like an umbrella was to be used to give the needed heat protection and to cool the cabin down to a liveable temperature. It worked! The flight crew then freed the jammed solar panel and the mission was underway. We found out that we could get along very well with the power generated by the one solar panel and Skylab ended up being successful beyond all expectations.

Conrad and his crew stayed up for 28 days. They came down and the next crew stayed up for 59 days. They came down and the third crew went up for 84 days.

The Skylab, which was 22 feet in diameter, was fairly big and roomy (See Figure 6). This was the first time that any of the astronauts in space had the privilege of a stove to prepare their meals. Previously, they just ate candy bars or squeezed food out of plastic bags. Much to the delight of the crew members, they had real Earth type food, including a freezer loaded with steaks and all sorts of other good things. In the galley, there was a table with three food preparation positions (See Figure 7). The food was packaged in containers which held the proper individually sized servings. The crew could select whatever entree they wanted, drop it in a tray on the table, punch a button and it would automatically be heated to the right temperature.

This was also the first time there were toilets. You have to know astronauts fairly well to know how important that is to them. The procedure of doing normal body functions in the Mercury or the Gemini capsule or even in the Apollo was not a pleasant affair. Everything was done with plastic bags and was not very neat. Skylab had a complete space toilet. It didn't look like the one you have at home, but it functioned in a similar

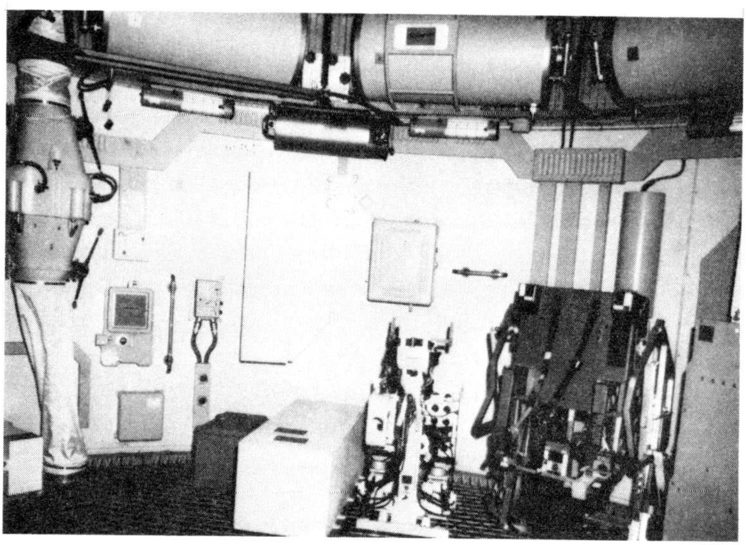

Figure 6

Figure 7

fashion, using air flow instead of water for flushing.

For exercise, the astronauts would run around the storage lockers that ringed the upper tank walls using them as a race track. They had an instrumented bicycle, called an ergometer. When they exercised they were wired up to get recordings on the condition of their health.

Other luxuries included a picture window for spectacular viewing of the Earth. The window was designed to be optically clear for picture taking, and was located in the galley where the men spent a lot of their time. They also had a recreation center with books, stereo and movies to make their stay more enjoyable. These are the things we must be sure to include when designing space stations for the future, where even longer stay-times will be required.

Skylab, with all that it did cost only about $2 billion over an 8-year period. Not an insignificant amount, but not large when compared to other government programs that do little to enhance our technical growth. It seems hard to justify cutting back our space programs when so much more money is being spent on other aspects of our economy.

NEW SPACE STATION STUDIES

The success of the Skylab program fostered new studies to take what we had learned and use it to design a bigger and more effective space station. This time, a 33-foot diameter space station was planned which could be launched on the huge Saturn V launch vehicle. This new laboratory was to be developed at the same time as the Space Shuttle so the two would go hand-in-hand. This project got as far as the "Phase B" stage where all the preliminary design work was done, preparatory to moving into the hardware development phase. Then NASA's budget was cut to the point where they could not afford both the Space Shuttle and the space station. They decided it would be wiser to build the transportation system first, to be followed by the space station which would be placed in orbit and serviced by the Shuttle. During this period, the Space Station Phase B Studies were re-oriented to a modular 14-foot diameter design seen in Figure 8. A full-size

Figure 8

14-foot diameter mockup was constructed and operational simulations were performed.

When Skylab systems were turned off in 1974, it was never expected that they would be used again. It was not designed to take on expendable resupplies such as control propellants, or the life support gasses of oxygen and nitrogen. Also, many of the scientists who had used Skylab for different observations decided they had collected so much data that it would be at least another five years before they could assimilate all the information that they had collected. Skylab was put to sleep, except for a radio receiver that could accept commands from the ground, which was kept active. After four years, we communicated with the receiver and found that it was working fine. We then started charging up the batteries using the solar panels and began turning on some of the other systems. The freezer, the environmental control system, the computer and the gyroscopes that stabilize Skylab all worked fine. This generated quite a bit of excitement because it looked as though the U.S.A. might still have a large space station available in orbit that could be visited by the Shuttle.

However, several miracles were needed before being able to use Skylab again. Initially, Skylab was launched at an altitude of 235 nautical miles (270 statute miles). The closer it got to Earth, the more drag there was on it. Even though there aren't very many air molecules up there, it's a large object and they eventually slowed it down enough so that it plunged back into our atmosphere. We wanted to get the Shuttle up to push Skylab back up into a higher orbit, but the Shuttle was not ready to fly on schedule with the necessary equipment to give Skylab a boost, and it was too late.

THE SHUTTLE AND SPACELAB

In the meantime, while the Space Shuttle is being developed, the U.S.A. studied the design for a sortie lab or what is now called Spacelab. This is a small space laboratory that fits in the cargo bay of the Shuttle (Figure 9). Meanwhile, the Europeans who had been trying to decide how they could join with

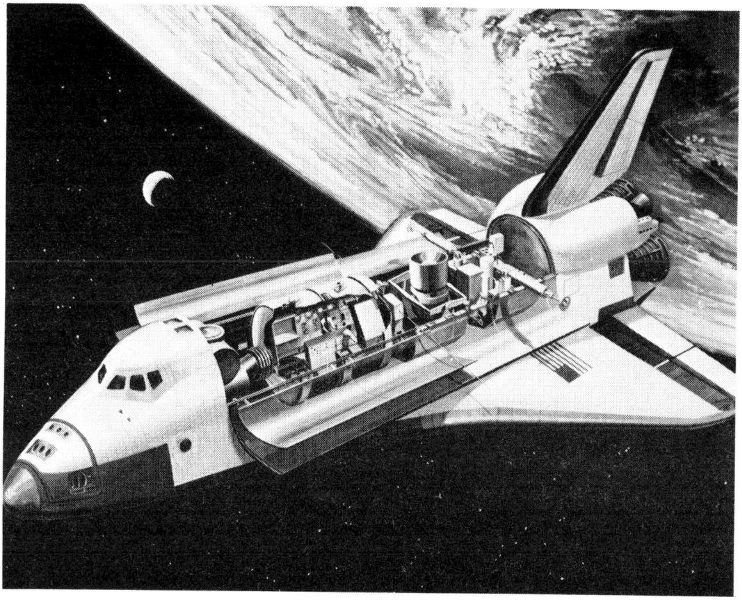

Figure 9

the U.S.A. in a joint venture in conjunction with the Space Shuttle, decided that they would finance development of the Spacelab to be used by the Western European Countries as well as the U.S.

The Spacelab design is a 14-foot diameter modular concept. A 14-foot diameter sounds quite small after Skylab's 22-foot diameter — but, as noted earlier, there is adequate room for people to move around and accomplish meaningful work (See Figures 10 and 11). NASA completed a full-size Spacelab mockup and ran a successful operational simulation test at JSC in Houston. A house-trailer was attached to simulate the forward Orbiter living area, where the crew ate and slept. The three-man crew ran operational experiments with rats, monkeys and plants — the kinds of things they would likely be doing in a Life Science Experiment Module. They ate the same food they would get in the Shuttle and stayed for seven days, which is the longest mission currently planned for the Shuttle. One of the key lessons learned was the operational

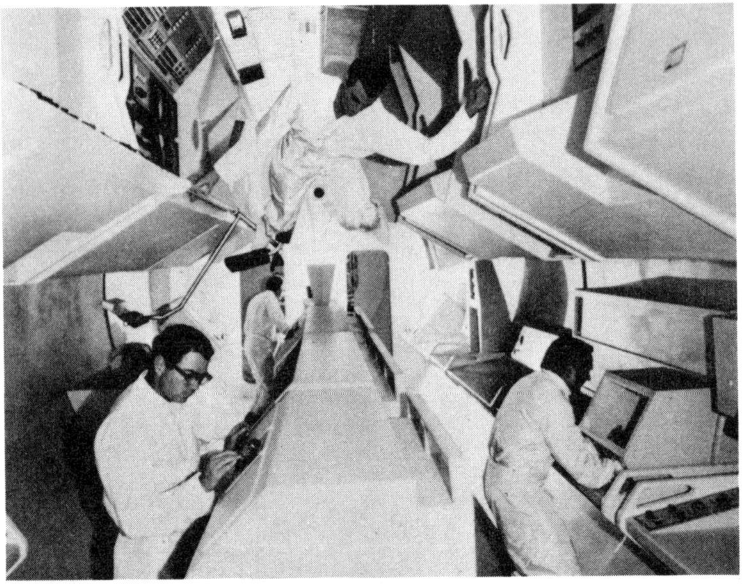

Figure 10

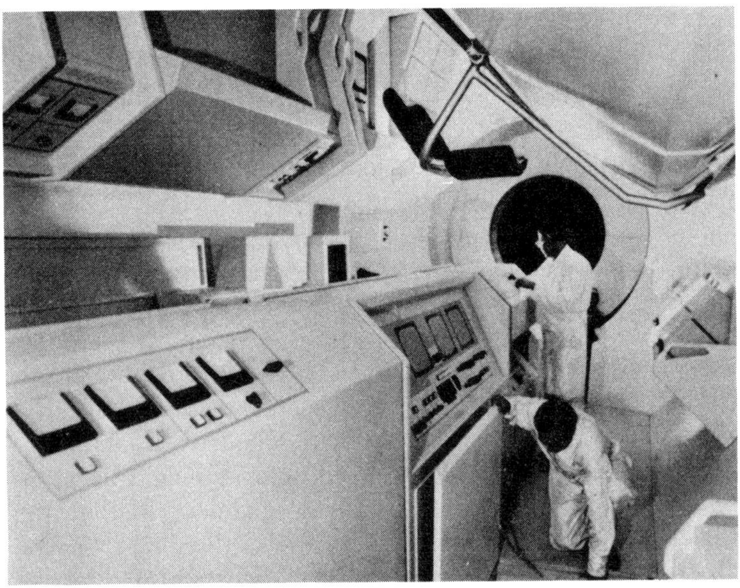

Figure 11

difficulty of the crew with mike cords becoming entangled while they moved about recording and conducting experiments.

The Spacelab pressurized module can also be used with an "open air" pallet. Experiments such as astronomy and infra-red sensors which need to be done with no obstructions to the space environment are conducted in the open bay of the Shuttle. The scientists running these "outdoor" experiments work inside the Spacelab or the Shuttle. Currently, the Spacelab is planned to be always in the cargo bay of the Shuttle, and therefore it is not a full-fledged space station. However, studies by both the U.S. and Europe are under way to use Spacelab components to provide a free-flying space station serviced by the Space Shuttle.

PLANNING FOR THE FUTURE

That brings us up to where we are today. Now I would like to discuss a little about the planning for the future, at least as I see it.

The Manned Orbital System Concept (MOSC) shown in Figure 12, was developed in 1975 using a double Spacelab module of the type the Europeans are building, together with a solar power module.

Figure 13 shows a four-man space station consisting of four basic modules. These modules are delivered to orbit in two Shuttle flights and when assembled become a space station of considerable capability. The Logistics Module that contains expendable supplies would likely be exchanged each time the crew was rotated — probably about every three months. The Experimental Module (Life Science Lab. shown) might also be exchanged.

Other experiment modules could be used for manufacturing, space processing or space construction, varying the content of Experiment Module to suit the desired mission.

This is an inexpensive evolutionary plan for an orbiting space station. The cost was kept low through use of existing

Figure 12

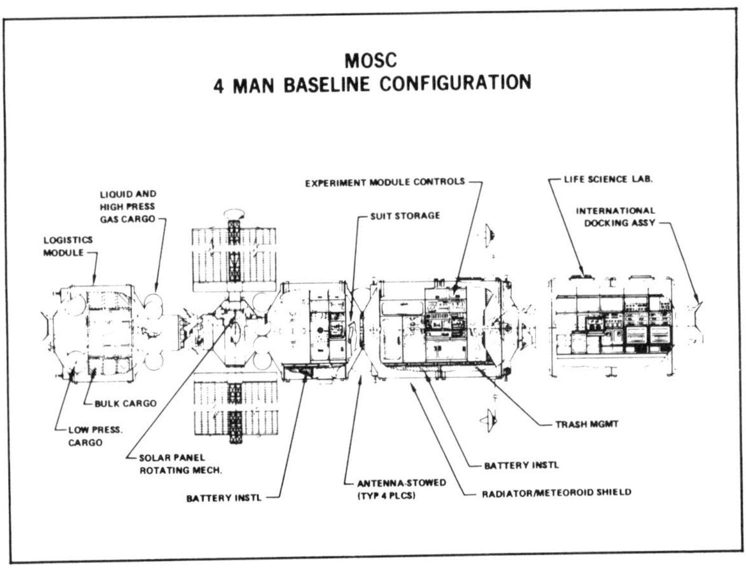

Figure 13

hardware from the European Spacelab Program and other Skylab and Space Shuttle qualified components. This modular space station idea of 1975 resulted in NASA's doing two new complete space station system analysis studies in 1977. Requirements for scientific research in Earth orbit haven't changed much in the ten plus years we have been looking at space stations. However, the new emphasis for the analysis studies was on the "Industrialization of Space," rather than on research and space exploration. In other words, the problem to be solved was: "How can we capitalize and exploit space for its unique advantages to the benefit of man?" This included ideas such as space processing of materials and pharmaceuticals, component manufacturing and demonstrations of new energy and communications systems.

SPACE INDUSTRIALIZATION

Two important space industrialization areas that provide service rather than products are communications and solar power satellites.

Communication Satellites:

In every neighborhood, every few square miles, we can see a telephone building which contains many kinds of switching equipment and relays. There are lines on poles or under the street with messages flowing from station-to-station. What we have discovered is that by putting the switching networks in space instead of on the ground, it will be possible to switch messages directly into our homes without having all the lines and other ground equipment. It is likely that there will be a transition period of many years where we will still utilize the current ground stations and lines. This is because of the hundreds of billions of dollars already invested on our communications networks. Such an investment cannot be discarded overnight. As time goes on and larger communication systems are built in space, a truly personal communication system much like the Dick Tracy wrist radio can become a reality. The technology to accomplish this is understood today, but requires antennas 100 to 300 feet in diameter in space at geosynchronous orbit.

Communication satellites have also spurred us into thinking about how such large antennas can be assembled in orbit. The Space Shuttle has a cargo bay about 14-feet in diameter by 55-feet long, hardly large enough for an antenna that is 100-feet or 300-feet in diameter. First, expandable or deployable structures that could be folded into the Shuttle cargo bay and unfolded in orbit were studied. These look feasible for smaller diameters, say up to about 30 to 50 feet. Later studies utilized man to assist in assembly of the structure in orbit. A problem with men working in space results from the inefficient use of their capabilities when they are enclosed in stiff pressure suits. Another consideration is that most of the astronauts to date have been test pilots, who want to fly the space machines, not become construction workers. The current breed of mission specialists are more interested in getting their experiments done. The kind of people needed for space construction are more like the workers we see putting up huge office buildings, bridges and towers. Perhaps all we have to do to attract these types of space workers is to offer "construction wages."

At any rate, we have concluded that to build large antennas there must be a compromise that provides for assembly of some critical components on the ground, probable fabrication of less complex structure with automatic machines in space and man-assisted assembly of major elements as shown in Figure 14.

Solar Power Satellites:

Another demanding requirement for large space structures is the Solar Power Satellite as illustrated in Figure 15, which shows a Space Construction Station which is building an early solar power test facility. A Solar Power Satellite collects the Sun's rays on photovoltaic solar cells, converts this into mi-

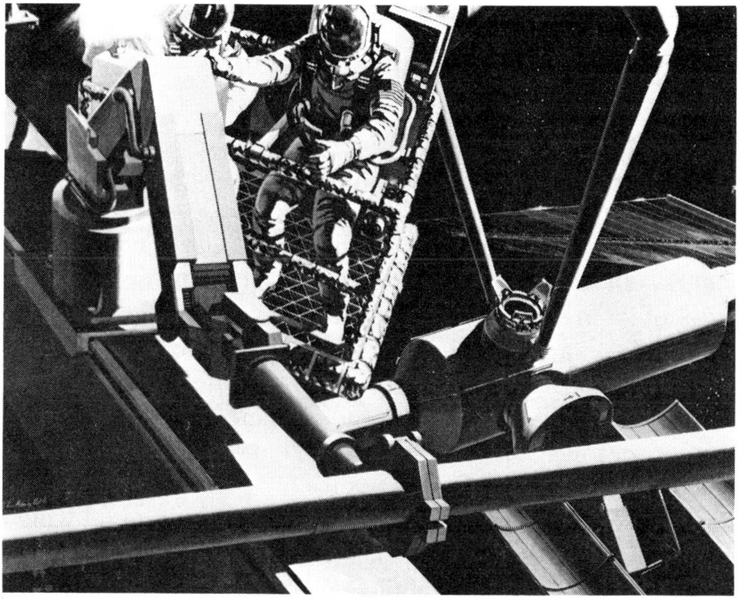

Figure 14

Figure 15

crowave energy and transmits it to the ground where it is received and fed into the electrical power grid system for distribution to homes and industry.

Solar Power Satellites are quite large, containing acres of solar cells. One 10 gigawatt satellite would be about three miles across by about thirteen miles long, weighing a hundred million or more pounds.

Some people are concerned about the safety and environmental effects of microwaving energy down to Earth. They think of microwave ovens and what they do to a hot dog in a matter of seconds. The question that always comes up is: "Suppose I'm flying across the country and pass through this microwave beam, will everybody in the airplane get roasted?" Proponents of solar power claim no problem exists because of the power level and very short exposure time. Another concern is that big receiving antennas are needed on the ground which create possible environmental problems in the surrounding area. "Suppose the rays coming down fly off and don't hit the antenna? Is that going to kill the people or animals who live in the area around it?" Again, the proponents say, "No." A sensor on the ground and another on the satellite will constantly and automatically correct the beam, keeping it confined to its proper path.

Another problem has been posed by some scientists who are concerned that such a concentration of energy will burn holes through the atmosphere. The best way to assuage all of these concerns will be to put a test satellite at geosynchronous orbit of sufficient size to prove that the concept really works and that the safety and environmental concerns can be evaluated.

The large Solar Power Satellites probably won't be built until well after the year 2000. Technically, they could be built sooner, but the impetus to do it is lacking. It is difficult to motivate the Congress to vote for the billions of dollars needed for the solar power program. However, when you consider the possibility that there might not be sufficient oil after the turn of the century, the Solar Power Satellite is a potential alternative. As a facility, it could be financed and amortized on the same 30-year basis that utility companies currently use.

Two countries appear more interested in solar power right now than the United States. One of them is England where they are just starting to collect money from the North Sea oil. They started a vigorous program developing North Sea oil because they couldn't afford expensive mid-east oil prices. Now they are starting to put some of the North Sea oil profits into developing new energy programs knowing that someday they will run out of oil in the North Sea. With a large Solar Power Satellite, they could sell electrical energy to other countries as well as being able to take care of their own needs.

Saudi Arabia is another country that has expressed interest in Solar Power Satellites. They also know that their oil will eventually be depleted. They are improving their standard of living very rapidly, and would like to maintain it in the future. Thus, they are interested in solar energy, not only for their own use, but also because they could save the oil reserves for their own use and export solar power to others.

Radiometer:

Another space construction effort that is being considered is a 30 meter (about 100 ft.) diameter Radiometer, to be used for making geological measurements, agricultural assessments, and ground water surveys. It can be assembled in space from a space construction platform shown in Figure 16. The Radiometer surface is made up of panels sized to fit into the cargo bay of the Shuttle. Surface accuracy requirements dictate that the panels be manufactured on the ground.

MINING THE MOON

It is one thing to say it is technically possible to build large Space Stations or huge satellites in space, but this leads to the obvious question of taking to orbit all the construction material from the surface of the Earth. It sounds like a monumental task when we talk of hundreds of millions of pounds. Further, it sounds very inefficient. There have been suggestions that we mine the Moon for the space construction materials. Pro-

Figure 16

fessor Gerry O'Neill, of Princeton University a few years ago unveiled a device called a Mass Driver that could propel materials from the surface of the Moon out to Lagrangian Point 5 (shown in Figure 17) which is a good place to collect and hold materials to build a space colony. Our geologists say the Moon contains the proper materials needed for such a project. The

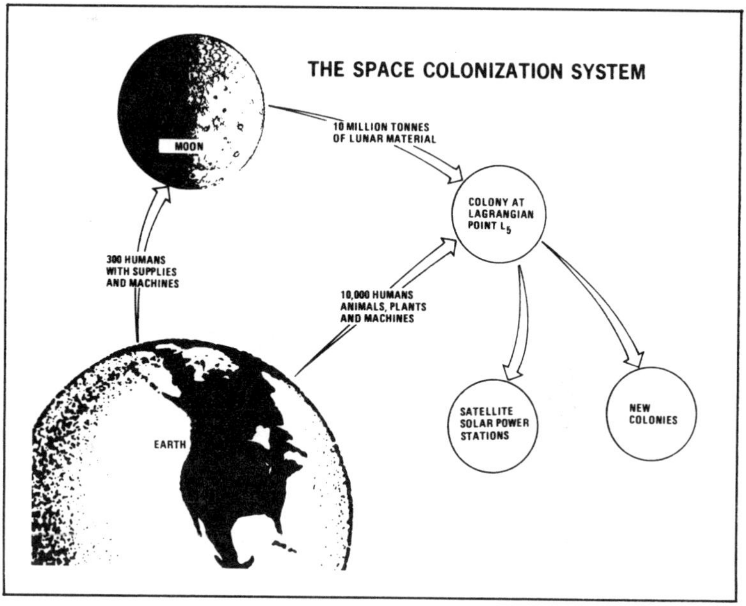

Figure 17

space colony inhabitants would then build Solar Power Satellites, again using lunar materials.

Figure 18 shows the design for a toroidal, or donut shaped, space colony. This habitat provides one "g" of gravity on the outer surfaces for a simulated Earth life environment.

INTERNATIONAL COOPERATION

There is an international flavor to all these plans. Many people think that these very large space endeavors will require international cooperation. It's just too big a venture for a single nation. The Europeans are already spending more than 500 million dollars of their own money developing the Spacelab for use with the U.S. Space Shuttle. This has saved NASA equivalent U.S. development dollars. It would have been difficult for us to have funded the development of the Spacelab and the Shuttle simultaneously. Thus we can expect that international cooperation is going to be more important in the future as

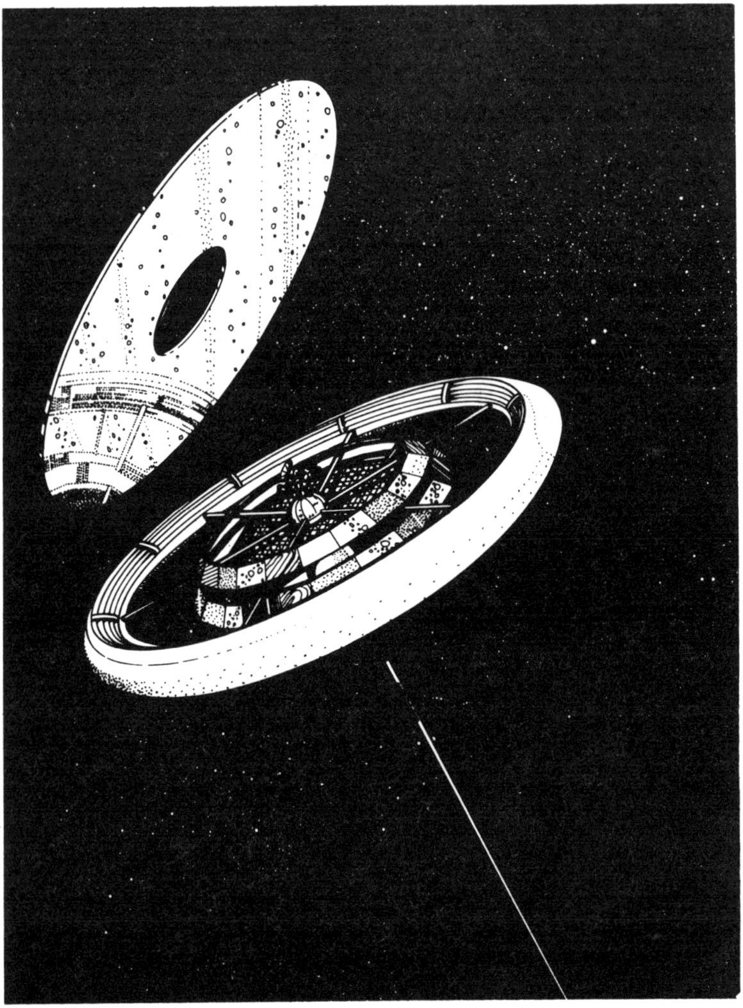

Figure 18

space projects increase in size and complexity.

Right now, the Japanese are making plans to use the Space Shuttle and the Spacelab for their national space research program, and there is talk of a cooperative effort between the U.S. and the Russians with the Shuttle docking in-orbit with the Russian Salyut space vehicle.

I don't want to leave you with the idea that we are going to have large space colonies for 10,000 or more people situated at L-5 in the very near future. There are many in-between steps that must be justified and accomplished. Large scale space projects will not be approved just because we have figured out technically how to do them. There must be a motivating force behind them and demonstrable results that demand the expenditure of these very significant resources. A much more likely near term plan will involve a low Earth orbit manned station capable of supporting space operations, assembly of space structures, materials and processing development, scientific evaluations and satellite services.

CHARLES L. GOULD

Manager,
Space Transportation System Utilization
Rockwell International

Charles L. Gould is Manager of Space Transportation System Utilization. Prior to that, he was the Chief Project Engineer, Advanced Programs and Technology, for Rockwell International's Space System Group where he had engineering responsibility for widely divergent projects, including manned orbital systems, new space communications and observation satellites, large space structures, and solar power satellites. He was also program manager of the study on Space Industrialization, the prospect of using space to produce goods and services of major economic benefit throughout the next 30 years.

Dr. Gould's aerospace background dates back to 1956, when his responsibilities at Wright Air Development Center related to the B-52 and the F-100 through F-108 series aircraft. Later, he was Engineering Officer at Ramstein Air Base, Germany. Since his Air Force duty, he has been involved in many advanced projects, first as a government employee and later (1962) at Rockwell International. During the 1960's he worked on both Apollo and Skylab and was manager of electrical/electronic systems for the Apollo modifications for Skylab. In the early 1970's, he was assistant chief engineer for the Space Station and advanced logistics system studies performed for NASA. Later, he was engineering manager for several military projects and managed his division's internal research and development program.

Dr. Gould holds a BS degree in mechanical engineering from Iowa State University, a certificate of business management from UCLA, a Ph.D. in business from California Western University, and is a Certified Professional Manager. He is also a member of Pi Tau Sigma, the national mechanical engineering honorary, and is active in the AIAA.

2

LARGE SCALE HUMAN BENEFITS OF SPACE INDUSTRIALIZATION

CHARLES L. GOULD

INTRODUCTION

Space industrialization can be defined as a new technology in which the special environmental properties of outer space are used for the social and economic benefit of the people of the Earth.

Although forecasting 30 years or more ahead is always suspect, some trends seem clear. One of these is that we will almost certainly experience a doubling or tripling of the world population in the next 100 years. Barring major catastrophies, more and more of the people on Earth will live in developing countries. These countries will control vast mineral resources and a huge potential pool of labor; but without the technology of a developed world, their full potential will not be realized, at least not within the realm of a developed world. Certainly it could not be realized within this century.

This relationship between the rich and poor nations is thus one of the major factors that is shaping our national as well as

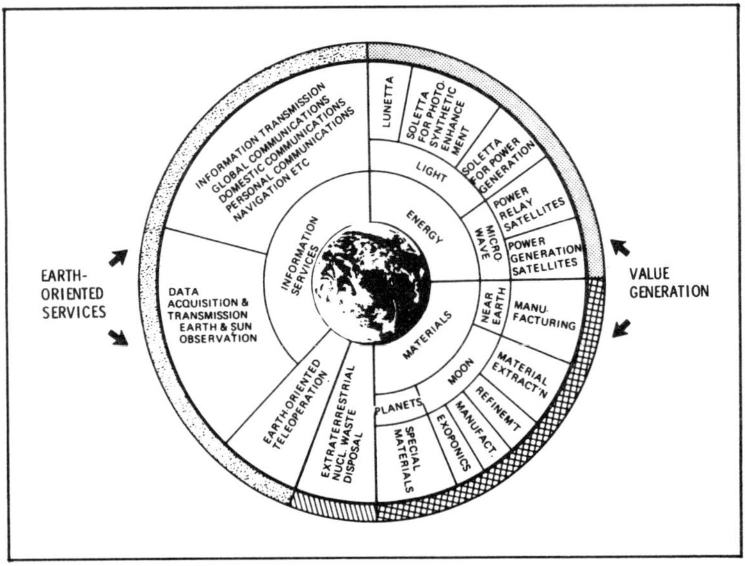

Figure 0

our global future. The industrial use of space can be of extreme importance to us as a nation, but it can also do much for the advancement of the developing countries throughout the rest of the world.

From September, 1976 to April, 1978, I was manager of a NASA Contract on Space Industrialization. The study had two parts: Part One identified the potential goals for space industrialization and looked at ways to realize those goals. Part Two was concerned with the programs and problems of actually bringing about the goals.

The study looked at parallel approaches simultaneously. One approach attempted to forecast general future trends. When we look at the last thirty years we know that there have been a lot of surprises; there will be even more in the next thirty years. Nevertheless some things are fundamental and we tried to identify those things in order to find out what the future needs of people worldwide might be. We also did technological studies, trying to determine what was actually possible and then to match that to the expected needs. We found literally hundreds of options and compared them to terrestrial solutions to see if space could offer unique, cheaper, or quicker solutions than terrestrial ones. Based on these results, we narrowed the possibilities down to a few that we called Anchor Opportunities (see Table 1). Then we designed the machinery that might be able to carry out these opportunities. It was not necessarily the best, but it was machinery that could do the job.

Space industrialization is confused in many peoples' minds. We define it in a very broad sense. You can use space to provide very large scale benefits either supported by the government in public services, supported by private industry or some combination thereof. In this definition of space industrialization you can divide space into three areas. The first, science and exploration, is basically what we have done already. It is one area of space that has many spin-offs, but it is not in-

COMMUNICATIONS	NAVIGATION, TRACKING, AND CONTROL
1. Information Relay • Direct TV broadcast • Electronic mail • Education broadcast • Rural TV • Meteorological information dissemination • Interagency data exchange • Electronic cottage industries • World medical advice center • Centralized "distributed" printing systems • Environmental information distribution • Time and frequency distribution **2. Personal Communications** • National information services • Personal communications wrist radio • Voting/polling wrist set • Diplomatic hot lines • 3-D holographic teleconferencing • Mobile communications relay • Amateur radio relay • "Telegraphing" personal communications systems • Worldwide electronic ping pong tournaments • Central computer service (for transmitting hand-held calculators) • Urban/police wrist radio **3. Disaster Warning** • Disaster warning relay • Pre-disaster data base (earthquake) • Earthquake fault measurements • Disaster communication set	**1. Navigation** • Public navigation system • Global position determination • Coastal navigation control • Global search and rescue locator **2. Tracking and Location** • Implanted sensor data collection • Wild animal/waterfowl surveillance • Marine animal migrations • Vehicular speed limit control • Rail anti-collision system • Nuclear fuel locator • • Vehicle/package locator **3. Traffic Control** • Multinational air traffic control radar • Surface ship tracking **4. Border Surveillance** • U.N. truce observation satellite • Border surveillance • Coastal anti-collision passive radar

Table 1-A

tended to directly benefit people like us. Next is the military area, and finally everything else comes under the term space industrialization. Space colonization is superimposed on all of these areas, as it is a step beyond and should not be confused with space industrialization.

LAND DATA

1. **Agricultural Measurements**
   - Soil type classification
   - Crop measurement
   - Crop damage assessment
   - Global wheat survey
   - Crop identification/survey
   - Agricultural land use patterns
   - Crop harvest monitor
   - Range land evaluation
   - Crop stress detection
   - Soil erosion measurement
   - Agricultural acreage survey
   - Soil moisture measurement
   - Soil temperature monitor
2. **Forest Management**
   - Timber site monitoring
   - Logging residue inventory
   - Forest stress detection
   - Forest fire detection
   - Rural/forest environment hazards
   - Lightning contact prediction/detection
3. **Hydrological Information System**
   - Snow moisture data collector
   - Wetlands monitor
   - Tidal patterns/flushing
   - Water management surveillance
   - Irrigation flow return
   - Runoff forecasting
   - Inland water/ice cover
   - Subsurface water monitor
   - Water resource mapping
   - Soil moisture data collector
   - Irrigation acreage measurement
   - Acquatic vegetation monitoring
   - Saline intrusion

3. **Hydrological Information System (cont)**
   - Underwater vegetation survey
   - Lake/river suspended solids
   - Sediment measurements (rivers)
   - Flooded area monitoring
4. **Land Management**
   - Land capability inventory
   - Land use mapping
   - Wild land classification
   - Range vegetation mapping
   - Rangeland utilization/population
   - Flood damage assessment
   - Beach erosion
5. **Pollution Data**
   - Advanced resources/pollution observatory
   - Salt accumulations (irrigation)
   - Agricultural pollutant monitoring
   - Lake eutrophication monitor
   - Great Lakes thermal mapping
   - Effluent discharge patterns
   - Toxic spill detector
   - Air quality profilometer
   - Air pollutant chemistry (freon)
   - Pollution detection and distribution
   - Mosquito control (wetlands flooding)
6. **Resource Measurements**
   - Oil/mineral location
   - Drilling/mining operations monitor
7. **Geographic Mapping**
   - Urban/suburban density
   - Recreation site planning
   - High resolution earth mapping radar
   - Wildland vegetation mapping
   - Offshore structure mapping

WEATHER DATA

- Atmospheric temperature profile sounder
- Rain monitor

OCEAN DATA

- Ocean resources and dynamics system
- Marine environment monitor
- Oil spill
- Shoreline ocean current monitor
- Algae bloom measurement

GLOBAL ENVIRONMENT

- Glacier movement
- Ozone layer replenishment/protection
- Highway/roadway environment impact
- Radiation budget observations
- Atmospheric composition
- Energy monitor: solar terrestrial observatory
- Tectonic plate observation

Table 1-B

FUTURE FUNDAMENTALS

In the Space Industrialization Study we attempted to look into the future at the fundamentals. Figure 1 shows the most basic consideration of all: population and the population distribution. The population explosion is real. We have done quite well using medicines to save the lives of people in developing countries. However, the average age in the developing world is about 15, or younger than students in a high school class. (Compare that to the average age of 29 in industrialized countries.) That means that if these persons live for three generations and only have two children per couple, the world's population will just about triple. And thus, regardless of birth control success, we are committed to supporting a large number of people in the future. We have about 4 billion people in the world today and the very lowest projections are 8 billion leveling off in a hundred years or less. The most probable projec-

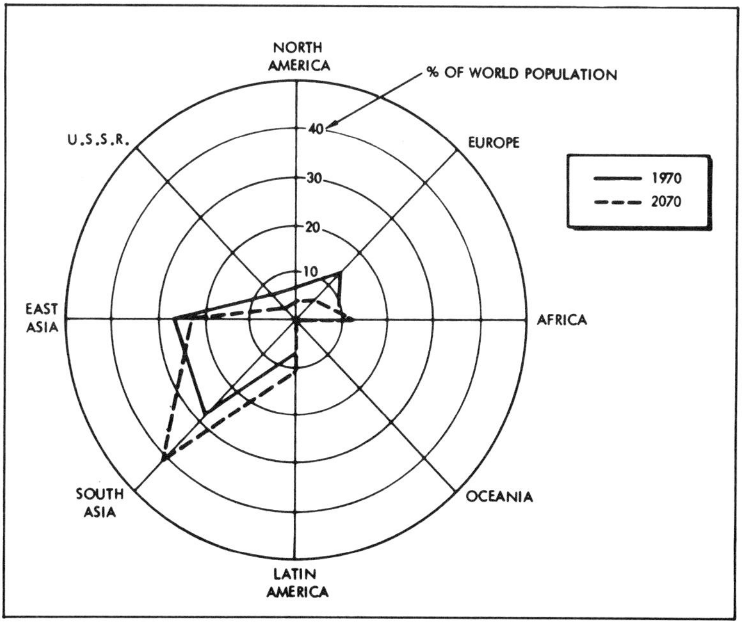

Figure 1

tions are perhaps twelve billion. Some people have said the figure may easily be twenty-five billion. That is an immense number of people and it is the overwhelming driver of what the world future is going to be like. We see that the industrialized world as we know it is not where all the people are. Even more significant, it is not where they are going to be in the future. The majority of the population is going to be in the developing countries where you have the greatest number of young people with a heritage of large families.

Some prestigious people got together an organization called the Club of Rome and funded some studies by MIT on future projections. These results had a great influence on many people in our political system, including Governor Brown. The result, a book by Dennis Meadows, *The Limits to Growth*, is fundamentally a doomsday prediction.

Now we have had doomsday predictions about population trends and resource limitations ever since Malthus, and they have not been proven true. Figure 2 is taken from that book. It gives the rate of usage in percentages of many interesting minerals and shows when we would run out of those minerals at reasonable prices or several times the reasonable price. It predicts we are going to run out of a lot of important minerals within our lifetime and the report concludes that we can not continue doing the kinds of things that we are doing and keep on going. "We must think small."

It may be true that we have supplies of minerals that we can get and could use, but the problem is that as we go deeper and deeper, it takes more energy, and if we go to lower and lower grades of ore, more energy is used. We may have sufficient aluminum and iron in our clay, but it takes a great deal of energy to dig it up and process it. Actually, everything we might do takes more energy than the current way of doing the same thing and that is why energy is of fundamental importance. If there is an energy shortage, there is going to be a major shortage of many other things. If there is no energy shortage, then we can attempt to get more of the minerals out of the lower grade sources. Thus we have two fundamental problems — population and energy.

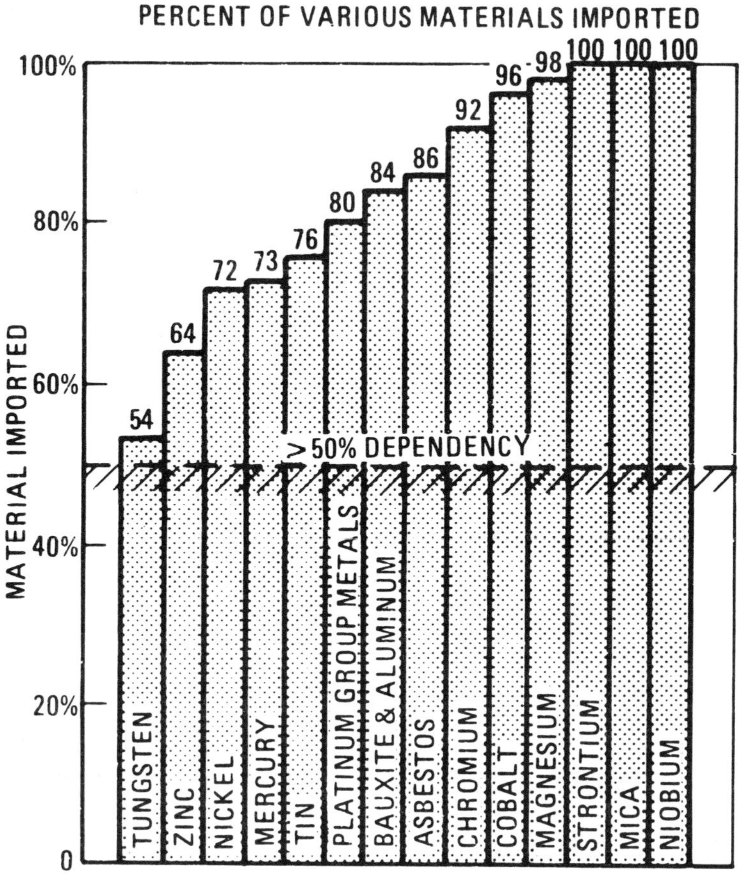

Figure 2

An energy shortage is not just a shortage. It is also a financial arrangement. Figure 3 shows our oil imports over the years. As far as the United States is concerned we insist on doing things the way we have always done them. We have not really cut back much and probably do not intend to. But the amount of money that flows out of the country is getting greater and greater. A dollar spent within the country rolls over two or three times and gets taxed each time. A dollar outside the country represents what would be termed by the stock market a "call against our resources." If we take all the resources of the

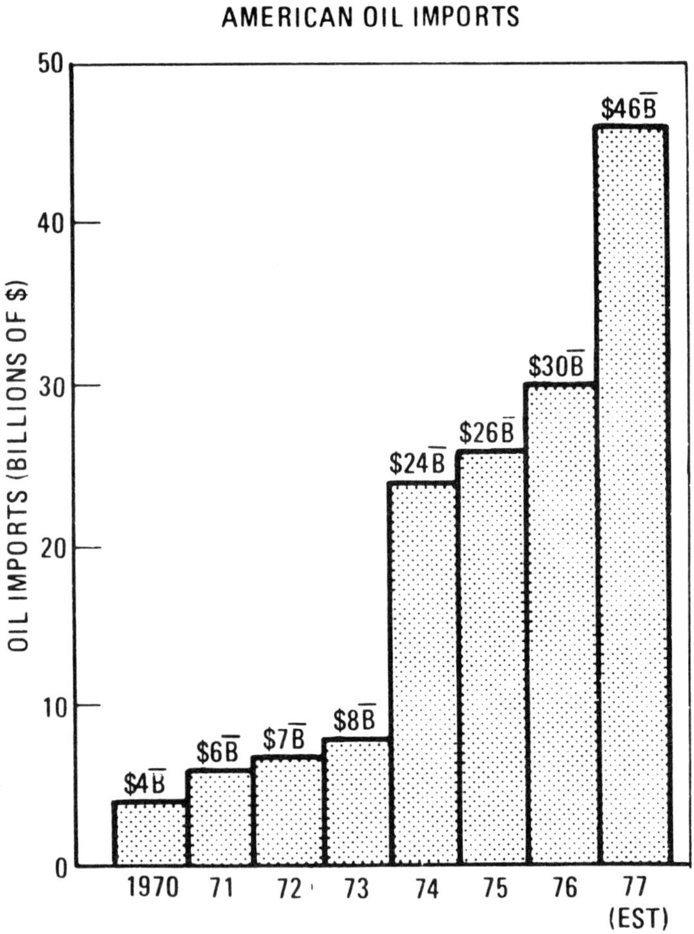

Figure 3

United States such as trees, buildings, roads, etc. and place the current stock market value on them, and then estimate all the amount of oil there is in the OPEC countries at current prices, the situation balances out. We are busily exchanging one for the other. The oil problem then, is a serious economic problem.

We are also dependent on other countries for many other things (see Figure 4), creating an economic situation representing pressures that can be applied to us in a political sense. Figure 5 is a chart made up by Dr. Krafft Ehricke that shows exports, more or less matched by imports, and population. We find that the overwhelming population is in the developing countries. If they had buying power they would represent a

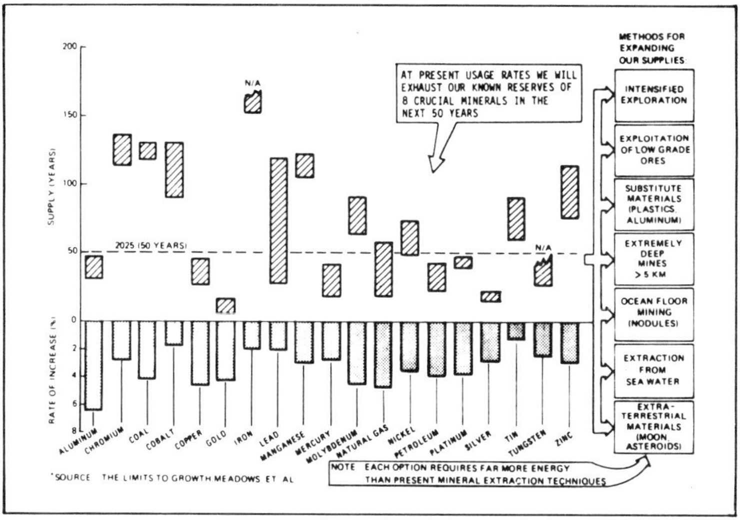

Figure 4

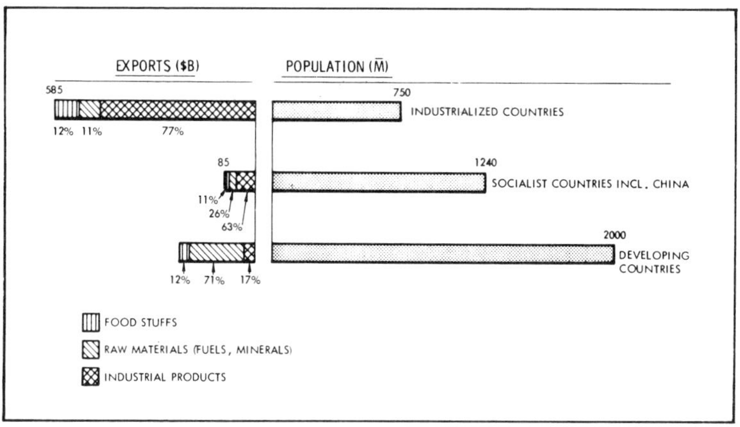

Figure 5

considerable market. Exporting raw materials gives them some power to buy the kind of things that we would like to sell them and that they would like to buy, including high technology products. But they don't have enough money to buy, and so they are not yet customers in the true sense. Therefore I think that the relationship between the industrialized countries and the developing countries is probably the most important thing that is going to affect us in the next thirty years. If this relationship goes sour, the prospects for a good world future for these billions of people, including ourselves, is not too promising. If this relationship is good, then I think that the future can be very hopeful. We have what they need, which is knowledge and industrialization, and the kinds of products that they need to bring themselves out of poverty. They have what we need, which is raw materials. If this relationship works right, the future is bright. If not, I think that we have a very dismal future.

One of our consultants, Dr. Reed Powell, put it quite succinctly when he told us that if we look for the quality of life directly, we are apt not to find it. We need to look at all three areas of energy, productivity and the balance of payments because, in the long run, the standard of living is based on them.

Dr. Bertrand Chatel, Chief of the Office of Science and Technology of the United Nations has addressed the issue of the needs of the developing countries. They are really quite different. We cannot say that one developing country is like another. We have to take them one at a time. To summarize the problems and the developing countries, we see that they have many problems and needs. Energy certainly is one of their needs, just as it is one of ours. But if we think about it, their costs for energy are higher than ours because they do not have the economics of scale we have and they do not have the transportation systems we have. In the middle of Bangladesh a gallon of gas is going to cost more in real terms than a gallon of gas here. Of course we all need the energy for the same reasons — to pump water out of wells, to power tractors, and manage resources.

In the U.S. when unemployment reaches six or eight percent, it is considered a serious problem. But unemployment runs sixty or seventy percent in developing countries generation after generation. Everybody needs a job in order to feel good about life so they can be productive and contribute to an integrated world. They need to do some of the manufacturing of their raw materials in their own country in order to have jobs and food.

Food is another problem. I think that it is possible to feed a population three or four times as great as the world population is now, but we have to do it cleverly. Food scarcity is an inhuman way to level population. Water of course is another problem. Frequently countries have water down a few hundred feet just as we do, but the people do not know where it is and they do not have any means to bring it up.

Perhaps the most important problem of all for developing countries is education. If people living in these countries could take that latent brain power and educate it, they would have a resource that will be very useful to the world. But it needs to be done in a few years. More education would secure for them better jobs through a policy of feedback. If they were more productive at the end of the year than they were at the beginning, they would start the next year hoping to improve. Then they would really be on the road to coping with their problems. Productivity needs to keep pace with the population growth; in fact, it needs to exceed the population growth.

Finally, the United States needs to maintain its leadership role. We have the technology and the economic power to make things happen in the world. In Krafft Ehricke's words, there were two ways of life — the creation of wealth versus the organization of scarcity. Conservation is a good thing. But in the case of increasing populations and increasing expectations it will be difficult to cope with the organization of scarcity. This situation could eventually lead to wars. It is a problem on which we need to put our brain power to work.

Many people would like to go into space just because it is fun to be there. But I am not enthusiastic about space colonization because, by our wildest imagination, it is not going to make any difference in the world population problem as it has been suggested in some early books on the subject. Colonization will come after we have made the technical ground work by doing things that pay for themselves. We don't risk a lot of money on something that is not proven. That is why I think the human activities in space will come after a lot of the productivity is developed, and eventually when we go back to the Moon.

ATTRACTIVE OPPORTUNITIES

Now to address the question of space. What does space have to offer? Space is a nice place in many ways, but it is not a panacea for everything. I happen to be in the space business and my topic is space; inevitably we'll hear a lot about that. But that does not mean that we should not do other good things like mine coal, develop terrestrial solar power, etc. The vantage point of space is very important for information transmission, observation, and the weightlessness, cleanliness and vacuum up there are important. The fact that you have sunlight all the time is important not only for space industries but perhaps for power on Earth.

We divided the opportunities to be gained from space into four categories; services, products, energy and human activities. As you see in Table 1, there are many attractive opportunities. For example, in the area of weather we already have weather satellites that are now quite useful. We watch the forecasts and satellite pictures on the television daily. We understand a great deal more about the weather now because of these satellites; however, we are not very good at predicting the weather a long way in advance yet. Such a skill could be very important for the management of resources and agriculture. Furthermore, if you had a climate change in Canada or South Dakota where for some reason the snow did not melt for most of the summer, because it might be reflecting so well that there would be a lack of the snow melting for many summers, something would need to be done. If we were able to see this coming by being able to predict the solar cycles, we could perhaps dig up some coal and make it into soot and spray the snow. Then the black soot would absorb the Sun's heat and melt the snow, thus getting the situation back to normal.

Information Services satellites are already making money. We are learning some very interesting things about the universe and the Earth as that technology gets better and better. For example, the Global Positioning System can give time to the accuracy of ten to the thirteenth and it can give a position with an accuracy better than thirty feet. A man or ship with a receiver gets a signal from the satellite, analyzes the information, and computes the latitude, longitude and height positions. These and other information satellites are now important factors in our civilian and military communications systems. They work quite well.

The reconnaissance satellites use the infrared spectrum and the focal plane of those satellites is very delicately designed and has to be kept cold. If these satellites don't work, the cornerstone of the SALT agreements with the Russians is taken away because we use those satellites to see that they are living up to their side of the bargain.

You could also have satellites automatically track icebergs and perhaps help tow them to areas that need water. The satellite in space would navigate and operate ocean tugs which use sails whenever the wind is blowing and then switch to diesel engines when it is not. The whole system could be unmanned.

Table 2 shows opportunities in the products or space industrialization area. One of the things I do not want you to do is confuse space industrialization with space manufacturing. Space manufacturing is just one of several things that we might do in space and in my mind it is probably the weakest near-term possibility. If you take all the experiments that have been done in space manufacturing, you could probably put them all in a briefcase. This refers to actual crystals that have been made, metals melted, and so forth. Yet if you take all of the books that we have written about those experiments they would fill up a whole telephone booth, and maybe more. We have talked too much and done too little. It is impossible to predict a potential product from such information, because of

A. PRODUCTS (REPRESENTATIVE)

1. ORGANIC

- ISOENZYMES
- GENETIC ENGINEERING OF HYBRID PLANTS
- UROKINASE
- INSULIN
- NEW ANTIBIOTICS VIA RAPID MUTATION
- UTILIZATION OF BIORHYTHMS

2. INORGANIC

- LARGE CRYSTALS
- SUPER LARGE SCALE INTEGRATED CIRCUITS
- TRANSPARENT OXIDE MATERIALS
- SURFACE ACOUSTIC WAVE DEVICES
- NEW GLASSES (INCLUDING FIBER OPTICS)
- TUNGSTEN X-RAY TARGET MATERIAL
- HOLLOW BALL BEARINGS
- HIGH TEMPERATURE TURBINE BLADES
- SEPARATION OF RADIO ISOTOPES
- HIGH STRENGTH PERMANENT MAGNETS
- MAGNETIC BUBBLE MEMORY CRYSTAL FILM
- THIN FILM ELECTRONIC DEVICES
- FILAMENTS FOR HIGH INTENSITY LAMPS
- ALUMINUM-LEAD LUBRICATED ALLOYS
- CONTINUOUS RIBBON CRYSTAL GROWTH

Table 2

the lack of concrete experimental data. We have made some lists of what would be possible, but I would not want to bet on the odds of any *one's* becoming a marketable product. However, the list is so long that the chances of *some* of it coming true are very high. When we have an environment with weightlessness, I am sure that we are going to have a lot of industries that will work best in that condition and it only takes a few breakthroughs to make a whole new industry. For example, look at the electronics industry.

However, we have some tough competition in this area from the Russians and the French. The French have been trying experiments on Russian space vehicles. The advantage would be to secure the first world-wide patent rights on some space product. Let us say that the research on turbine blades works out and they get a patent on it. Then they can sell high temperature turbine blades to the world. Most of the nations cooperate with patent laws. Thus the French would then have a patent and we would have to pay royalties if we wanted to use the product.

In the area of space processing we publish the results of our experiments completely. If the Russians have done anything at all that is successful, they are ahead of us because they have all our reports. They do have a furnace in their Salyut-6 space station that operates at about 2200° which is extremely hot and larger than any furnace that we have ever planned so far for our Spacelab.

We also have the possibility of generating energy in space (see Table 3). It is a good place to reject heat and it is a good place to do fusion research. Space is a place that can be used for disposing of waste heat with no threat of pollution. (The Sun pollutes the solar system all the time.) People have also talked about putting nuclear waste into space and there would be no

B. ENERGY

1. LUNETTA
   - NIGHT ILLUMINATION FOR URBAN AREAS
   - NIGHT ILLUMINATION FOR AGRICULTURE AND INDUSTRIAL OPERATIONS
   - NIGHT ILLUMINATION FOR DIASTER RELIEF OPERATIONS
2. SOLETTA
   - NIGHT FROST DAMAGE PROTECTION
   - LOCAL CLIMATE MANIPULATION
   - REFLECTED LIGHT FOR GROUND ELECTRICITY CONVERSION
   - OCEAN CELL WARMING FOR CLIMATE CONTROL
   - CONTROLLED SNOW-PACK MELTING
   - STIMULATION OF PHOTOSYNTHESIS PROCESS
3. OTHER
   - SATELLITE POWER SYSTEM (SOLAR)
   - FUSION IN SPACE
   - NUCLEAR WASTE DISPOSAL

Table 3

need to worry about polluting the Earth. From this, however, I think that we ought to put it in a place where we can go get it again. I think that my grandchildren will be smarter than I am and perhaps those wastes may turn out to be valuable materials. There have been many things that people have thought were unusable waste that turned out to be quite valuable.

If we are going to build a lot of large cities and if we really want to light those cities at night to prevent crime, etc. we could do the whole job from space with no cables or lights. We would reflect sunlight directly down to the spot we want to light by means of sodium covered reflector satellites. We could also have a power generation satellite that would be used to turn sunlight into electricity that could be sent through existing power systems.

There is also a possibility of getting raw materials fom extraterrestrial sources. There are asteroids that might be very usefil. For example, perhaps we could put a small solar powered thruster on an asteroid about the size of Texas that is pure nickel and change its orbit so that in a few hundred years it would float by the Earth to be retrieved.

ANCHOR OPPORTUNITIES

With all these possibilities we found, the trick was to narrow the ideas down to a few. There is no way that you can throw out two hundred ideas to the general public and have them be anything other than confused. We wanted about 25 or 30 "Anchor Opportunities." To get them, we used several different criteria adding up to some five thousand decisions that had to be made.

Some things need to be near-term that could be done quickly, because congressmen get re-elected every two years and Presidents don't like to go out on a limb on something that is not going to pay off within their term of office. This is also true of industry officials who are anxious to get on with it. If it is near-term and important, then it is seen as good by both groups. If it is long term, even though it is very important, like energy, it may have a rough time. This is one of the Achilles' heels of Solar Power Satellites which probably will not happen until after the turn of the century, and there is a lot happening now and before then in the energy business. However, both information and observation services are very near-term as well as quite important, and thus our study sees them as winners in the first steps of space industrialization.

The next task in developing this list of "Anchor Opportunities" was to ask questions of a fundamental nature, such as "What are we trying to do?" and "What could sell?" The result was six options. One was to respond to current needs or what people perceive to be their needs because that is what is real and that is what can get funded. Anything else will not get much of an audience. Another option is to plan for a farsighted space program based on the idea that the future belongs to those who prepare for it. If we really have the foresight to hurt a little today for a better tomorrow, our chances are much greater at having a better tomorrow. Foresight then should be the basis of the space program. If the space program can postpone World War III by even a day, it would pay for itself.

Space, furthermore, is by its nature international. It is hard to go into space without somebody else knowing you are there. The Spacelab has been constructed by the European Space Agency in cooperation with various European nations and the U.S., showing that we can work together. Space is the only frontier that man has ever had that is only two hundred miles away from everybody, and people do look to space as a new frontier.

There is a lot of money to be made by all in the space power and information markets, and this could be one way the U.S. could obtain money to balance our international payment situation. We in the high technology community have an obligation to point the way for the future and use some foresight in planning the space program. The spirit and direction can be pulled by the long term, but any program must be responsive to the immediate needs of people. Otherwise, it will never be funded.

Based on that narrowing down process, we developed the

final Anchor Opportunities, shown in Tables 4-7. The "R" stands for "research", "D" stands for "development", and "O" means it is operational. The primes and double primes indicate a change in the operational status. An arrow represents a saturation of the market. Finally, these Anchor

Anchor Opportunities	Time Frame				
	89-85	85-90	90-95	95-00	00-10
	Services				
Transmission					
Direct-broadcast education-U.S.	0	0′	0″	→	→
Direct-broadcast education-Developing countries	D	0	→	0′	0″
Business system data transfer	0	0′	0″	→	→
Electronic telecommuting	R	D	0	→	→
Electronic teleconferencing	D	0	→	→	0′
World medical advice center	D	0	0′	→	→
Time and navigation services	0	→	→	→	→
Implanted sensor data collector	D	0	0′	0″	→
National information services			D	0	→
Personal communications	D	0	0′	→	→
Electronic mail (excluding packages)	D	0	0′	0″	→
Medical aid and information - U.S.	D	0	→	→	→
Teleoperation from space			R	D	0
Observation					
Oil/mineral location	0	→	→	0′	→
Crop measurement	D	0	0′	→	→
Ocean resources and dynamic system	D	0	→	→	→
Water resource map and runoff forecast	0	→	→	→	→
Global effects monitoring (STO)	0	0	0′	0″	→
Landsat D	0	→	→	→	→
Topographic mapping	D	0	→	→	→
High-resolution resource survey	D	0	→	→	→
High-resolution radar mapping	R	D	0	→	→
Microwave radiometer		0	0	→	→

D = Development, 0 = Operational, R = Research
0′ or 0″ = Step increase in capability
0 ——→ = Plateau of activity

Table 4

Anchor Opportunities	Time Frame				
	80-85	85-90	90-95	95-00	00-10
	Products				
Organic					
Isozymes (also medical diagnostic)	R	0	→	→	→
Urokinase (anticoagulant)	D	0	→	→	→
Insulin (from human sources)	D	0	→	→	→
Inorganic					
Large crystals (size and perfection)	0	→	→	→	→
Super-large-scale integrated circuits	0	0′	→	0″	→
New glasses (including fiber optics)	D	0	0′	0″	→
High-temperature turbine blades		D	0	→	→
High-strength permanent magnets	0	→	→	→	→
Cutting tools	D	0	→	0′	→
Thin-film electronic devices	D	0	0′	→	→
Continuous ribbon crystal growth	D	0	→	→	→

Table 5

Anchor Opportunities	Time Frame 80-85	85-90	90-95	95-00	00-10
Energy					
Reflected Solar Energy					
Night illumination for urban areas	R	D	O	O'	→
Night illumination for agricultural and industrial operations			O	→	→
Night frost damage protection			R	D	O
Reflected light for ground-electrical conversion		R	R/D	D	D
Microwave Transmission					
Satellite power system (solar)	R	R/D	D	O	O'
Fusion in space	R	R	D	D	O

Table 6

Human Activities					
Medical and genetic research	D	O	→	→	→
Space vacation cruises (Shuttle flights)			D	O	→
Orbital tourism (LEO Hotel)			R	D	O
Orbital Therapeutics				D	O
Entertainment and arts			O	→	→
Lunar					
Unmanned explorers	D	O	→	→	→
Lunar orbiter		D	O	→	→
Lunar base		R	D	O	→
Lunar industry			R	D	O

Table 7

Opportunities are the ones that we think are the best out of the 200 possibilities.

SERVICES

To show you a specific case, let us take one option in the Services Area, namely the Implanted Sensor Data Collector. A large antenna would be on orbit in space and you would have a small transmitter on Earth about the size of a coffee cup. The transmitter would give its serial number and beep its signal which would be received by the orbiting antenna and routed to an Earth receiver. For example, this product could be used as a burglar alarm that would go off if tripped, sending a signal to the police. In another situation, a transmitter on your electric meter would measure your time-of-day-use of electricity and then adjust your bill accordingly. This would tend to level out peak loads relative to the base load of available energy. This would provide the means of averting brownouts and blackouts in times of high use. Such units would cost about ten dollars.

Furthermore, this is something that can be done in a reasonable time period.

Looking at the basic idea behind the area of Services, I would like to discuss the concept I coined called "Complexity Inversion" (See Figure 6). In the past in the manned space program we were willing to construct things of almost any size on the ground just so that the receivers in space could be very small and light as in Apollo. All of the large computers were on the ground so the Apollo electronics could be very simple, and for a good reason. We did not want to make the Apollo capsule too heavy, and we wanted to make it extremely reliable. Since then we have found electronics have a long reliable life in space which has proven to be a good place to store and operate such things. Most satellites last longer than we want them to last, and in fact are cluttering up things up there. We have to turn them off even though they are still operating. When the Space Shuttle gets going and we have manned flights going into space, people will be able to modify things in unpredictable ways and this will allow us to put large things in space thus making it possible for its complement on the ground to be very small. The antenna on the ground can be the size of a coffee cup if the antenna in space is big enough. The power

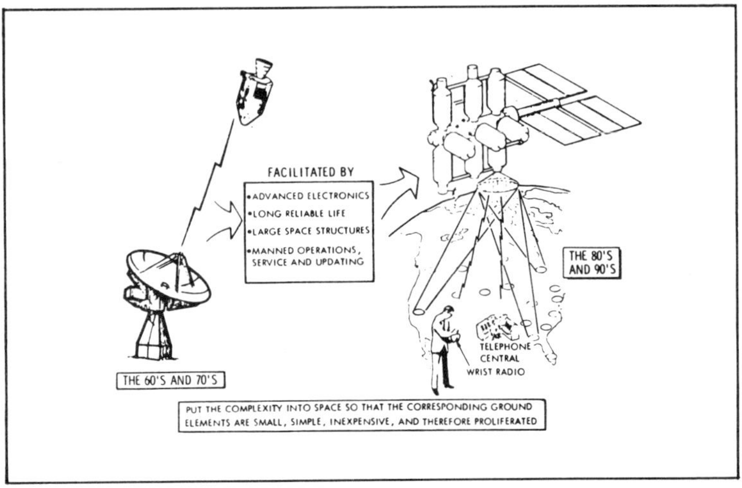

Figure 6

needed on the ground can be almost nothing in terms of watts if the antenna in space is big enough and pointed accurately enough so it will pick up a very tiny signal. And the key to this is cost. If it is cheap enough then we can have billions and billions of units on the ground. Thus we have Complexity Inversion. The complexity we had with Apollo is inverted so that the simple units are on the ground and the large complex unit is in space.

One example of Complexity Inversion would be Telecommuting. We have done a lot in this country to conquer our geographical limitations. Geography has been an important matter. It has involved culture as well as transportation. However, space can take the importance of geography out of the picture through communications that will provide direct broadcast no matter where you are, and independent of local facilities. A person could work at the New York Stock Exchange without driving down to that little island every day in the snow. Certain kinds of services could be done from almost anywhere. The Los Angeles School District at one time had all their key punch operations done in Taiwan. Even though they transported the tapes by plane, it was still a type of electronic shipping that bypassed geography, leaving it unimportant. With space Complexity Inversion by means of satellites we could live anywhere in the world we desired, and have many of the benefits of a large urban area like that of Los Angeles. There would be access to schools and education of all kinds that would be very cheap. In fact, this type of Complexity Inversion has proven so cheap that the underground and undersea cable industry have had to lower their prices in order to stay competitive. If we had access to the Library of Congress through a computer terminal it would not matter if we lived in a small town in Kentucky with no library. This opens up vast new jobs. If everybody has access to a whole nation via satellite someone can become a specialist in something like locating parts for the '57 Chevy and start up a business by computer. If someone wanted to be a key punch operator, the key punch machine could be operated from any place. The same is true for selling insurance or managing insurance forms. These functions could be done from any place.

World markets can also be opened up through space. In the less developed countries education is very important, and teachers in short supply. Much of that problem could be alleviated by the use of satellite broadcast systems. We need to do a lot of research yet, but the experiments done so far have been to show that people can learn to interact with television education without knowing how to read or write. In that case we could reinstitute something like an apprentice system. An apprentice system took a one-to-one relation between the master and the apprentice. The apprentice did not have to learn a lot of reading or writing to become a good wood carver or craftsman. Television can use that same method to teach those in the developing countries because you can tailor the material to their specific situation. Even though we are broadcasting billions of things, a specific set would only receive on a prescribed beam. You can turn the set to a particular station, being very selective in developing the right course for a given group of people, suited to the job that they can do next year. This might be a skill like basket weaving or making hats that a tourist might buy. The system can work if it is cheap enough.

On the ATS 6 in India we found out that this system worked quite well. But ATS 6 was an experimental satellite with a short life span. It is about to die, and India is now trying to buy their own satellite for education.

We can also use the satellite to bring the outside world into classrooms in all parts of the world by showing major events or featuring speakers. The teacher can have specialized education on television screens by means of two-way access to people or computers which is a tool many local schools could not afford, especially in some of the less populated areas of the country. There can be programs for special kinds of people like the deaf or handicapped who can not leave their homes.

Eighty percent of small businesses fail in their first five years, not because they do not know the business, but because they do not know accounting, marketing, advertising, or how to fill out some forms. Access to educational programs and computer services could make a big difference to them.

In another kind of application, pocket telephones could become quite an interesting little device. A pocket telephone would be about the size of a small calculator. Some say it ought to be a wrist radio and in fact we have some already designed by Bulova. But I think that the first step really ought to be the pocket telephone because it is a little bigger and the batteries are a problem to get into a small wrist radio. These are quite feasible and we could have them now. Frankly, they would help the developing countries more than they would help us. We have a good phone system, but they do not. But they could have a good phone system, basically overnight, by putting up an antenna in space that would connect an entire continent with pocket telephones. A paramedic who has been trained to do simple tests could go into the countryside, and make analyses that could be discussed with physicians and specialists anywhere in the world if necessary. Many people in the developing countries have been considered lazy, but now we find out that they are not lazy at all. For generations they have battled internal parasites so that their energy is gone. A good paramedic could go to these village people with a minimal education and could be in contact with specialists on parasites in Chicago, Northridge, Bangladesh, or anywhere, for the same price. If you are beaming a signal out 22,000 miles it really does not make a lot of difference whether you are going to talk to someone a few hundred miles away, or another thousand or so around the Earth. The use of pocket telephones in developing countries would be a way for us to help those people without trying to impose our values and our social structure on them. We would be providing them the means to help themselves. From a safety standpoint, we could always use a pocket telephone to call the police or report an accident at some remote location. These units could also be programmed to do certain things or appear to have a certain recording in them. If one has a heart problem, the recording would say: "I have a pill in my pocket here. Take the pill out of my pocket and stick it in my mouth." The unit would also be a calculator, and it would have access to all the big computers you wanted through space.

ENERGY

Next let us consider the list of energy products opportunities. We do not know which will be a good investment yet. A few years ago the National Academy of Sciences came out with a report that said that there is no way to predict that there is going to be a big energy industry in space. We are still in a research development phase. I certainly agree with that. That is why we are building the Shuttle and Spacelab. We want to get up there with people and do research and development to find out what works. Shortly we will be able to use the Shuttle and we will have people up in space for seven days. If an experiment does not work right, we can take it back and forth in no time. However, the Shuttle is power limited. I think that we will have to put extra solar cells on it so it can stay on orbit at least 30 days as well as have a lot more power to use on the experiments. Many of the projects we would like to do take an oven which requires not a few watts, but kilowatts of power.

At some point we will need a permanently manned facility with separate modules for experiments and materials processing related to the production of energy systems. If the decision is made really to go into production, it will require a lot of power. Next we will need to build space factories. If that works out, we will have more than one and they will become larger and larger. We will probably have that sort of thing about the turn of the century.

Let us look at the energy situation from a turn of the century standpoint. The needs for energy are going to increase and the concerns include the cost, the politics worldwide, the environmental impact, and the institutional problems of who will control things, who will run things, and who will make a profit.

It is true that we ought to be more efficient than we are, but that is not going to solve the problem. It is just going to ease it a bit. We will probably use more hydrocarbon fuels in the way of coal in the future, but believe me, coal is dirty stuff. There is no way that coal is going to be nice and clean, and we are going to have to tear up a lot of land to get that coal out in a cost-effective manner. When we burn coal, no matter how many controls we have, it is going to put wastes into our very fragile and very thin biosphere. Terrestrial solar systems will help and we are working hard and doing some good things in that area. But it takes a lot of energy to make a solar cell, and if it is ten years before we have paid back that energy, we have not gained much. There is a lot of research going on in that area but I think it is not going to solve the problem.

Fusion, energy of course, is an enormous thing. If it comes along I do not think that we will need SPS until we get production thermal barriers. If we are doing things with more and more energy, we are wasting heat. Practically every process that we have is a heat process in the energy scheme, and eventually when we use the energy, we produce heat. In a television set, the waste product is heat. There are a lot of questions being raised about the abilities of the Earth's biosphere to contain all that heat.

As we become more and more industrialized, we may hit a thermal barrier somewhere. If the ice caps melt, and the oceans rise a hundred feet, it would affect a large number of cities in a very drastic way, and Howard Hughes was right in buying Las Vegas because he wanted a seaport. We really do not want to change the temperature of the Earth much, but there is a possibility that in the next hundred years we could surpass the Earth's thermal tolerance.

That is why fusion in space is an interesting possibility. All the heat is dissipated in space. Krafft Ehricke thinks that there is a good possibility for the breakthrough in nuclear fusion research to come in space because we can move the container walls further away from the fusion reaction and we can also restore a vacuum very quickly whenever impurities develop. Therefore the superconduction magnetic bottle works very quickly.

Once we have fusion in space there are two options: one is to leave the reactor in space and beam the power down to Earth. The other option is to do enough of the experimental research in space to ensure confidence in the process and given enough of an investment potential, build a sufficiently large vacuum chamber on the ground. To understand the significance of a fusion breakthrough, try to imagine the ocean as gasoline and multiply it times three. That is how much energy there is in fusion.

The objective now with Solar Power Satellites (SPS)(Fig. 7) is to find out whether or not we really have that option. If we do, it would be clean, redundant energy we can beam down in very efficient ways, and it can be converted to electricity at about 85% efficiency as compared to 15% efficiency achieved using solar cells on Earth under a noonday Sun. Again, we leave most of the heat that is rejected in the process in space. To decide whether or not SPS is feasible, many technologies must be proven. SPS is a huge structure even though it is very light. The paintings of how a SPS may look are nice to look at, but in a sense represent the "battle of the air brushes." The real challenge is to get the costs within reason and to find out if it can do what is expected or not. How much will it weigh? How reliable will it be? How long will it last? What effect will be associated with such a narrow beam very concentrated in one frequency? The beam will no doubt have some harmonic frequencies and side lobes. The area above the rectenna would warm you up. You might think birds would not want to go there, but they may fly there too much in the winter time. We have to make sure that there are no effects in the various layers of the ionosphere, and that there are no far reaching spill overs. If the rectenna is very efficient in picking up the waves sent down we can go ahead and grow crops and so forth underneath. But that has to be checked out carefully. This careful checking is part of the reason for the expense and the long duration of the program so far. There might be major effects that might be unknown or unpredictable which could cancel the whole project if not detected early enough. Some people are afraid that the microwaves will hurt us as they come down. At the ground level the beam has less energy than ordinary sunlight. So the amount of energy that is coming down through these microwave beams is really nothing compared to the Sun. The Sun puts out energy on almost every wave length. In fact the Sun itself poses many problems. If you have the antenna in space, it is very hard to get a signal through the Sun's output of energy because the Sun simply overwhelms everything. That is one reason why Dr. Ehricke favors the use of reflectors such

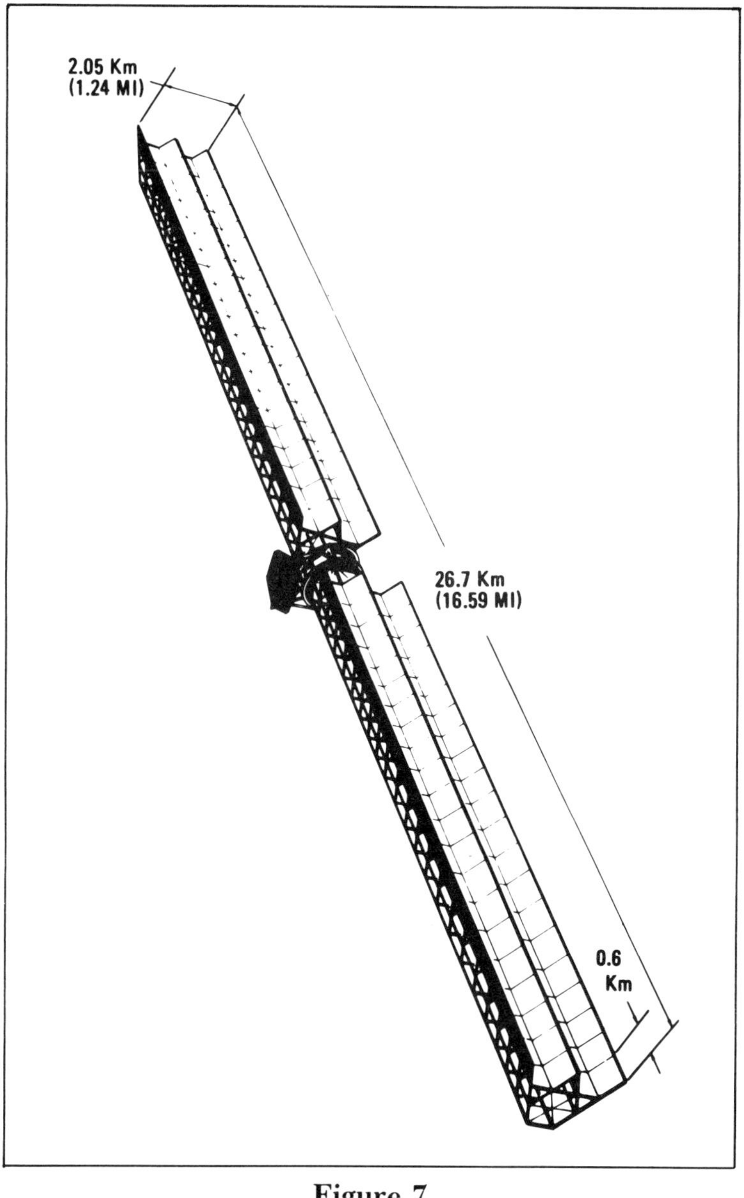

Figure 7

as "Soletta" and "Lunetta."

A solar reflector such as Lunetta or Powersoletta would be a large disk covered with sodium which simply reflects the sunlight down to the Earth. Because of the size of the Sun and the laws of optics, there is no way such a beam could ever be focused to a small pinpoint of dangerous light. The beams would be fairly large, and a series of reflectors in a low orbit could light huge areas at night, or concentrate the light for smaller areas that are paying for the service. If we wanted to light a city the size of New York, rather than have one very bright beam that makes shadows all around, it would be better to have about a half a dozen reflectors which are all moving over the city so there are no shadows. We can almost read with one full moon of light on a clear night. The Lunettas could create about 50-100 moons of light depending on the weather. Lunettas could be used for many things. In developing countries they could help to harvest precious crops at night, thus creating the food production and providing jobs.

Lunetta, however, does not produce much energy, only light. A Powersoletta would produce energy. The reflector would send beams of sunlight down to large solar collectors on the Earth's deserts both night and day. Thus large useless areas of desert could in fact be converted into energy production. Another important use for the Powersoletta would be in frost protection. If we had a reflector that was, for the most part, dedicated to providing power for Barstow, and we had a killer frost that we could predict in Florida, all we would have to do is divert the Powersoletta for one or two nights in order to raise the temperature a very small amount. It would not be enough to upset the ecology, but it could save the entire fruit crop, representing billions of dollars that could be saved.

Another use for the reflector is called the Biosoletta which is used to produce more food per acre than any other scheme I have ever seen. There are certain places in the ocean where you have "upwelling" regions. Here the nutrients rise to the surface of the ocean and only need a small amount of light to develop the tiny plant called plankton which is the basic food of small fish. Most of the time this area is covered by clouds, and thus the transformation process does not take place. With the Biosoletta, however, you can shine light on these areas of the Arctic oceans and produce more protein than the best farm in Iowa — enough to feed about 2 billion people a day, simply by increasing the fish supply chain.

The study ended with recommendations for the three decades of the 80's, 90's and beyond 2000. (See Figure 8). What we recommended for the first decade are the weather satellites, information and observation satellites, and power sources. There should be a space station at a low inclination orbit. The satellites in geosynchronous orbit will eventually become so numerous that it will make sense to put several together on one large platform. This might provide some five hundred kilowatts of power for pocket telephone relays and direct broadcasting channels for education and telecommuting.

In the next decade we would add some free flying modules for space manufacturing that are visited once every six months. In the 90's we see research in space on nuclear fusion and the very first SPS could come as early as the turn of the century. In the last decade we have a mature space program. It has everything in it. In this case you either have many SPS's, or if SPS does not work out, you would be reflecting solar energy down or you may have fusion in space. Sooner or later one of these options is going to be completely renewable, providing Earth with essentially infinite sources of energy. You will have a lot going on in low Earth orbit and many "buses" going up and down. People could go there for a vacation because there will probably be tourist facilities by that time. You will need to bring many materials from the Moon primarily because of the activities in geosynchronous orbit. It is much easier to get them from the Moon than from the Earth because of the Earth's deep gravity well. The transportation after the turn of the century will be basically a shuttle, along with unmanned boosters for cargo and better inter-orbit transportation using solar electric motors for slow transfer and chemical propellants for faster flights that are manned. By then we will probably have a bigger and fancier ship called a Starraker (Figure 9) which will be a huge single-stage-to-orbit plane that could take

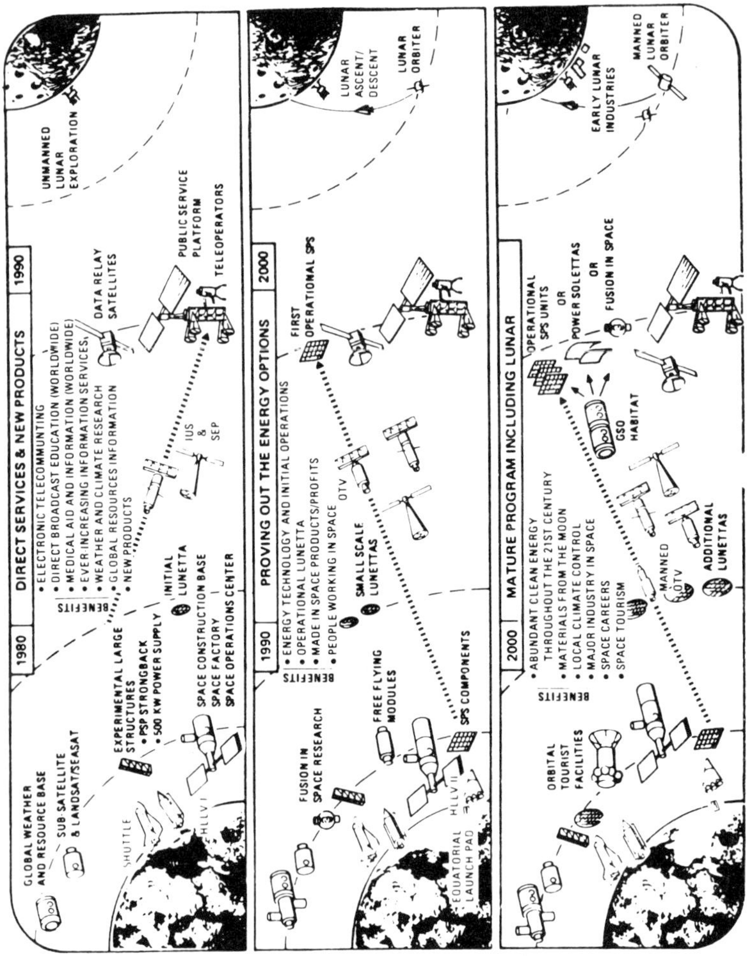

Figure 8

off from Los Angeles Airport like a current jet, go into orbit, and land again like a conventional airplane.

CONCLUSIONS

Stewardship is essential and this country needs to learn that lesson a lot more than some of the other countries. We are not

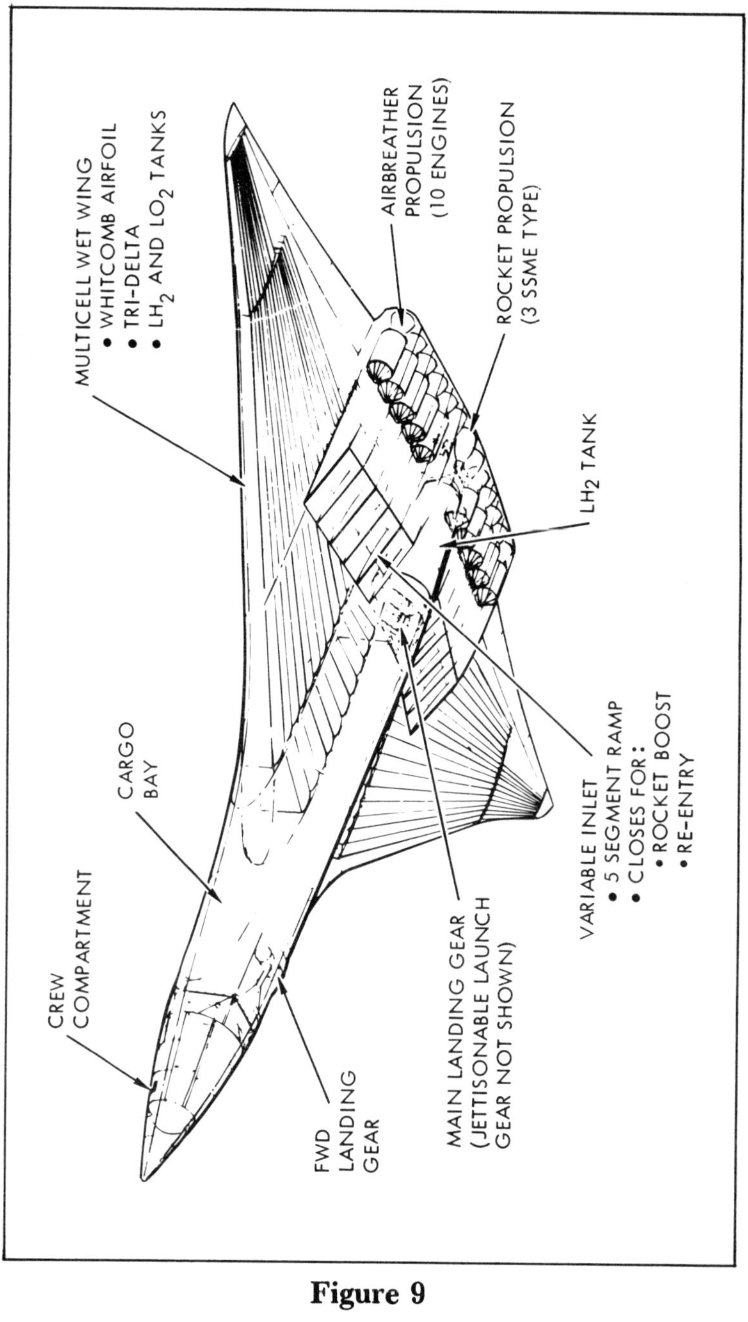
MULTICELL WET WING
• WHITCOMB AIRFOIL
• TRI-DELTA
• LH_2 AND LO_2 TANKS
AIRBREATHER PROPULSION (10 ENGINES)
ROCKET PROPULSION (3 SSME TYPE)
LH_2 TANK
CARGO BAY
CREW COMPARTMENT
VARIABLE INLET
• 5 SEGMENT RAMP
• CLOSES FOR:
• ROCKET BOOST
• RE-ENTRY
MAIN LANDING GEAR (JETTISONABLE LAUNCH GEAR NOT SHOWN)
FWD LANDING GEAR

Figure 9

very proficient in taking care of our resources. Eventually we have to discover something that is renewable like solar energy. I think that the critical time period for mankind is within our lifetime. We have the capabilities to destroy the Earth if we want to. The former generations couldn't. If we do the right thing within our lifetime, then I think things will be in a more stable situation and there will be greater understanding of the fact that we are all in this world together. If we do not do the right things within our lifetime, I think mankind as we know mankind, won't be able to survive, and that is quite a responsibility to place on people like us.

It behooves all of us to have a reasonably happy world population. It turns out that where people have jobs to do and reasonable expectancy of happiness in their old age, they have fewer children and that eventually tends toward stability. If there are jobs in the developing countries, then there are markets for the kinds of things that American industry makes and sells. For example, Rockwell manufactures electric meters. It would be nice to have an additional demand for electric meters from about 3 billion people. If they had electricity, there would be 3 billion more electric meters that somebody could sell, and that would be good for business. We like to have customers that want to buy our products, but if we have someone that is almost starving to death and does not have a dime, that does not help the stockholders at Rockwell very much. They would rather have such a person with buying power and the productive capabilities to make things that we will buy.

Our study clearly indicates that the 1980's will be the opportune time to bring space into our social and economic mainstream. Space industrialization should not be viewed as a competitor to science and exploration; rather, the vigorous program we have outlined can be the main impetus for expanding the scope of science and exploration. Likewise, the proposed program is not in conflict with the increasingly prevalent dream of young people to travel into space. Indeed that dream will inevitably become more credible and attractive in the next few years as the Space Industrialization program develops and evolves into a mature undertaking.

FAREN R. AKINS, PH.D.

Research Associate,
University of Santa Clara

Dr. Faren Akins is currently working as a Research Associate at the University of Santa Clara under funding provided by NASA Ames Research Center. For the past three years he has been completing a large mission planning project examining the role of the social sciences in the future of the space program. Previous to this he received a National Research Council Fellowship to conduct research at NASA Ames Research Center exploring the effects of isolation and confinement on non-human primate behavior. He has also received grants from the National Institute of Mental Health and from the National Institute of Health.

Dr. Akins received his Bachelor of Arts from the University of Tulsa and his Ph.D. in Psychology from the University of Arizona, Tucson. He graduated Magna Cum Laude, is a member of Phi Beta Kappa, and received many national and local awards, honors, and scholarships.

3 PSYCHOLOGICAL AND PSYCHOPHYSIOLOGICAL EFFECTS OF LONG DURATION SPACE FLIGHT

FAREN R. AKINS

I would like to talk with you today about some of the current behavioral research related to long duration space travel and thus to future space colonization. I would also like to share some of my views on what problems face us in our attempts to establish communities in space.

Based on our current knowledge of how people behave in other isolated unusual environments (Antarctica, submarines, etc.) it is my opinion that the notion of having an actual viable space colony in the near future is somewhat unrealistic. While engineers such as Gerard O'Neill, author of *The High Frontier*, have confidently stated that we currently possess the technical skills to accomplish such a feat, I am not at all sure that we have the necessary social science skills. There are a considerable number of problems which we have just begun to understand. Disregard for the problems of human interaction in isolated, confined environments could prove disastrous. At present, it would be quite difficult to select, train, and sustain an effective group of astronauts (or space colonists) for a long period of time in such an extreme environment as space.

One place we must start in our thinking is to define the goals of a space colonization mission. The operation of the mission, the leadership and social structure to be used, the psychosocial issues involved, and the eventual outcome of the project will all depend in part on what the specific purpose of the mission will be. Do we want to send people for economic reasons, for increased military defense, for scientific research, or for some combination? Actually, it is most likely that it will be for all these reasons. This tells us that the type of individuals which will be needed in space will be highly diverse ranging from blue collar worker, to scientist, to managers and supervisors. We can certainly anticipate from this that there will be problems in how such different people with such different values and purposes for being there will relate to each other and to the situation. It will be essential that we clearly define our mission goals and the roles and responsibilities of each project member. We can draw upon research in the polar regions to understand why this is so important. In these environments where scientists, military personnel, and non-military technical personnel have lived together there do appear to be some subgroup schisms. Each group has a slightly different view of the mission and the importance of their role in it. This sometimes causes conflicts between group roles which can be associated with frustration, decreases in self-esteem and potential conflicts between individuals and groups. From this we can conclude that future long term space missions will have to be very clearly delineated. Each person's role in this project will have to be clearly outlined and each role given importance whether it be custodian or unit supervisor, if people are to feel satisfied with themselves and their place in the overall project.

Given that a highly diverse group of people will be needed in space this also raises the issue of how the leadership and social structure will be defined or evolve. Some authors have suggested that an emergent form of government or leadership would be most suitable because it would permit maximum involvement of the colonists in the decision making process of determining how leadership will be defined. This is certainly a much different approach than the traditional military, chain-of-command model we tended to use in our space program to date. Probably something in between would be more suitable. It will be important to have a well defined formal system of organization and leadership well in advance of the mission. However, informal lines of communication will inevitably develop just as they do in any other type of group or work setting. This will be needed and should be encouraged to permit people to air their grievances and to feel more in control of the environment that controls their lives and work.

Just from this simple discussion it is clear that many questions must be addressed if long duration space travel or colonization is to become a reality. Let me discuss some of the other general categories of problems we need to consider.

BIOMEDICAL LIMITATIONS

Probably the single most limiting factor in terms of actually having a viable space colony or a long duration mission at this point, beyond the psychological difficulties, are some strictly biomedical limitations. The three month data that we have received from Skylab indicate very strongly that there are a number of problems involved in being in a weightless condition for long periods. Let us consider three critical biological systems that are directly and adversely affected by weightlessness: the cardiovascular system, the musculoskeletal system, and the vestibular system.

Without going into a lot of technicalities on what happens to the cardiovascular system when it is exposed to zero gravity, let me simplify by first stating the obvious: on Earth the heart is accustomed to pumping against gravity. However, in a weightless condition there is no gravity against which the heart has to pump. Consequently the heart slows down in response to decreased demands. As a result certain problems occur, such as a drop in blood pressure and volume, decreases in heart rate, and a shift in the body's fluid volume and concentration. Also, since the heart is a muscle like any other muscle, when it is not exercised it begins to decrease in size. Some astronauts have shown definite dimunition of heart size based on comparisons of x-rays before and after flight. Further, there have been some isolated cases of cardiac-arrhythmia in astronauts in which the heartbeat was very irregular and caused a certain amount of alarm among the medical personnel.

Another problem is the consequence of living in weightlessness when a human returns to an Earth-like gravity. The heart has been used to pumping at a slower pace, blood pressure has been reduced, etc. Therefore, when a human re-enters Earth gravity there may be difficulties in readapting. The heart must immediately react to this increased stress. Consequently, a greater burden is placed on the system and a period of readaptation is required. The stress of one gravity could be dramatic following very long term exposure to zero gravity if appropriate countermeasures are not developed and employed both during and after flight.

One of the countermeasures we have relied upon in the past is vigorous exercise. However, there is a growing concern that exercise may not really be effective in the treatment of weightlessness induced cardiovascular problems. Other techniques are now being investigated as alternative countermeasures. One of these is the use of lower body negative pressure (LBNP). Applying a negative pressure to the lower extremities tends to force blood back up to the upper part of the body and causes the heart to work harder. Tests so far conducted using this procedure have been encouraging. While this method and others we have used may be effective for short term flights, we are still not completely certain how the body will respond under long term conditions.

The system that currently poses the greatest difficulties for a long duration mission is the musculoskeletal system. Because there is no stress (weight-bearing) being placed against the muscles while in zero gravity, it is as if one never exercised. Without proper countermeasures the muscles begin to atrophy. That is the reason why exercise has been such an important part of the astronaut program. Our concern at this point however is what degree of physical conditioning we should require of our astronauts and what type and amount of exercise should be used. Originally, we tended to believe that highly athletic crewmembers were desirable. Superb physical conditioning was thought to increase the general resistance of the astronaut to the conditions of space. However, there is now a growing controversy that athletes may actually be less resistant to the adverse effects of weightlessness than non-athletes. Research has shown that athletes compared with non-athletes do not tolerate certain environmental stresses as well. For example, the aerobic (processing of oxygen) work capacity of athletes decreases at a faster rate during altitude increments than is the case with non-athletes. Athletes seem to tolerate such conditions as acceleration, centrifugation, LBNP, water immersion, etc. less well than non-athletes. Also, when exercise is used as a countermeasure to reduce the effects of these conditions it appears to be more effective with non-athletes than with endurance sport trained athletes. This certainly raises many questions regarding what type of exercise should be used in our space program and the degree of physical training our astronauts should receive. I expect that this will be an important research area over the next few years, particularly as we look toward longer term missions.

Looking now at the skeletal system we can anticipate major research efforts to investigate how zero gravity affects bones and how the adverse effects can be reduced. Because there is no stress placed on the skeleton, a process of decalcification develops. That is the bones begin to lose calcium that is not replaced. The loss rate is so low (about .3 or .4 percent a month was found during the 59 day Skylab) that no real problems occur during brief missions. However, we do not know at what point this process stops. During a very long mission it could continue for several years as it does with paralyzed subjects. This could lead to greater problems of fracture and also increased susceptibility to bone disorders and dysfunctions. Currently, we do not know of any truly effective countermeasures for preventing this effect. However, there has been some very interesting and encouraging research conducted in this area. One of the techniques NASA scientists are exploring is to place isolated bones under different kinds of stress, for example, mechanical or electrical. While this research is still in the preliminary stages it may eventually give us some clues on how to prevent bone degeneration in space. Another interesting research approach has been the use of inflight centrifugation. The skeletal system of rats subjected to centrifugation during one of the Soviet Cosmos missions appeared to have a positive affect. Bone mechanical properties including torque, stiffness, and energy were shown to significantly decrease among non-centrifuged flight animals, but were essentially normal among centrifuged flight subjects. While centrifugation did not correct loss of bone formation during flight, it did speed up recovery following return to Earth. Probably more research of this nature will be completed during our future space missions.

The third major system we have to consider is the vestibular system since it is involved with what has been called space sickness. The vestibular system very simply is your balance system — the part of your physiology that is telling you where you are standing, the position of your body, etc. What happens under weightless conditions is that receptors in the inner ear that normally rest in a fluid of sorts are affected by weightlessness and do not send out appropriate impulses regarding the position and relative location of the body. You get one set of information concerning your spatial position from your eyes and conflicting information from your vestibular senses. This sensory conflict, in some way which we do not understand at this time, produces the symptoms of nausea, vomiting, sweating, stomach awareness, etc. that have been reported by many of the astronauts. These symptoms are quite similar to those we associate with motion sickness on Earth. It is thought that similar mechanisms may be involved although the process is perhaps different. Fortunately, this is not something that is a real concern for a long term flight. The astronauts have been able to adapt to this condition in roughly 3 to 5 days. It only becomes a problem if critical tasks must be accomplished during the first few days of the mission. There is some evidence to suggest that it will become more of a problem as the size and variety of our space craft increases. It appears that a separate adaptation process is required in each new and different stimulus situation. Thus, while the astronauts of Skylab adapted first to the command module some of them had to make a second adaptation upon entering the workshop area. In some cases, astronauts who became ill during the first phase of the mission and then adapted to that condition satisfactorily became ill again upon entering the new surroundings of the workshop and had to make a separate adaptation to this situation. It appears that our previous learning and experience with perceived motion and perceptual habits greatly affects us in the non-gravity conditions of space. There have been a number of models and techniques that are being developed to deal with this.

One of the most interesting approaches is the use of biofeedback to counter the process of space sickness. Basically, the idea is to train subjects to better control certain autonomic functions which coincide with motion sickness symptoms. Through such control, subjects have demonstrated that they can tolerate much more intense levels of stimulation that normally elicit sickness with much lower incidences of illness. This is a fascinating application of behavioral techniques and a good example of how behavior scientists are working with certain biomedical problems using psychological approaches.

Now that you have at least a sketchy outline of some of the biomedical problems of humans in space, let us move on to some of the psychophysiological issues and then to the more psychosocial topics.

CIRCADIAN RHYTHMS

Most of you know that you are either a morning person, an afternoon person, or a night person. If you are like me you need your quart of coffee before the world looks a little better and you are ready to get going for the day. All of us go through some normal daily rhythms that are very well observed in terms of all kinds of physiological measures. One's metabolism, brain wave patterns, the amount of body heat being generated, blood circulation, heart rate, respiration rate — all of these things are subject to a normal daily cycle with peaks and valleys. The important thing about this is that these physiological circadian rhythms are based on certain cues or signals called "zeitgabers" that help regulate these functions. Obviously the greatest cue is the time of day based on the rising and setting of the sun. Our work schedule is another. Different social factors also influence our rhythms on a day to day basis.

In dealing with these rhythms during a space flight, we can run into some difficulties. How can one artificially reproduce Earth-like cues for time regulation? Without such cues our body does not regulate in a normal fashion and we experience many of the same effects we know on Earth as jet lag. This also relates to work schedules. Because of the often hectic schedules of the astronauts they have been prone to fatigue and desynchronosis (desynchronization of circadian rhythms). One area of concern for future missions will be to develop effective work-rest cycles to reduce the problems of circadian rhythm irregularities. The need for research to establish effective environmental "zeitgabers" will be important too. Some researchers have suggested using automatic lighting systems to simulate seasonal changes and daily illumination changes as they would appear on Earth. Other avenues might involve exploring how crews function on daily schedules other than 24 hours. We will also need to investigate how to most effectively arrange work schedules for rotating crews in space.

Related to the issue of circadian rhythms and desynchronosis is the problem of sleep in space. The time, amount, and quality of sleep is intimately tied to our circadian processes. The astronauts have reported some difficulties in falling asleep, remaining asleep and feeling refreshed and alert upon waking. Part of this has been due to the noise and vibration levels aboard the craft. However, it is unclear at present how weightlessness per se affects sleep. We will need more research to better determine how sleep in zero gravity is different from that on Earth and what effects this may have upon physiology and behavior.

ISOLATION AND CONFINEMENT

One of the biggest problems facing crews in space is that even the most motivated people who go into an isolation situation frequently become apathetic and bored if meaningful work is not required. While the work of crews in space will certainly be diverse and rigorous to begin with, there will be some problems as the mission progresses. As the work in space becomes more repetitive and as crews tend to specialize more in certain types of tasks the possibility exists that professional "burn-out" may occur. That is, people may become unmotivated by their work and consequently get caught up in the relative monotony of space. It will be important that we take sufficient measures to ensure that work in space continues to be meaningful. Certain prospects suggest themselves. For one thing, it may be possible to do some of the actual mission training in space during flight. Instead of completely training the crew on every aspect of the mission, perhaps some training can be done in space so that new challenges will be available. The work to be performed each day may be varied as well. Instead of scheduling the same tasks in the same order it may be possible to give individual crew members more freedom to select what work and in what sequence it will be carried out. These working conditions and the importance of work will be critical in the survival of crews during long term missions.

One of the very strong generalities that has come out of research on groups in isolation is that lack of meaningful work, the relative monotony of the environment, and the stresses of close, continued interpersonal contact can lead to increased psychosomatic complaints. That is, psychological stress can produce actual physiological changes including such symptoms as migraine headaches, problems with weight, mood swings, and depression. We must begin to develop more active means of dealing with the stresses of isolation through reward and motivational systems to help reduce these problems.

Knowing how long one is to be isolated and what the conditions of the isolation will be are very important. It is not advisable that crews be subject to indeterminate service duty. This is illustrated by one experiment in which subjects were confined in isolation and given misleading information regarding their length of stay. One group was told they would be there for 14 days, while the other was told they would be observed for 28 days. At the end of 14 days the first group was informed that a mistake had been made in scheduling and that they would actually be there another week. At the end of 21 days the second group was also told that a mistake had been made and that their time had been reduced to 21 days instead of 28. While both groups actually stayed in isolation the same period of time, the group given the underestimate showed considerably greater problems with depression, anxiety, and psychological stress.

While we have much more to understand about how groups in isolation function, we must also consider how groups in isolation relate to outsiders or to other groups outside their own system. For example, some research with isolated groups indicates that they tend to direct much of their hostility and aggression toward outsiders. Crews in the Antarctic are reportedly very wary of visiting reporters, etc. In the Sealab and Tektite missions where crews lived and worked on the ocean floor, there was a tendency to direct hostilities to persons in authority on the surface — through communication systems — linking the two. The amount of open hostility among group members in isolation is usually low. This may represent an acknowledgement that everyone must live and work together and that open hostilities led to ostracism and further isolation. As a result the direction of hostilities toward outsiders may be a way of allowing individuals to vent their frustrations without directly alienating themselves or other members of the immediate group. One could imagine that such problems in space could become disastrous. Should conflicts develop between mission control and the crew of a spacecraft the consequences for mission success could be serious. Probably the single most important person in the entire communication system of a long duration will be the person who is on the radio at ground control. Previously this has often been a fellow astronaut that is respected by the in-flight crew. As flights become longer and more detailed, we will certainly want to better understand the type of interactions that can be expected to occur between mission crews and ground control.

Another aspect of the interaction between crew and outsiders is what develops when rotational crews are employed. In the event that an orbiting laboratory becomes a reality the most logical way to staff such a facility would be with rotating crews. We then face the situation of bringing in outsiders and trying to integrate them into an already formed, working group. The person who is coming in may be looked upon as an outsider, someone who is not really a member of the group. This may result in skepticism about their abilities, both professional and social. There is often the expectation that they will do things in the same way as the person they are replacing. What effect will such a situation have upon the cohesiveness of the group and upon the functioning of the new group member? Such questions have yet to be adequately addressed.

There are many pitfalls in the isolation and confinement research so far conducted. Some of these limitations clearly influence what we can predict about groups in space. Future research efforts will need to consider these issues. For example, most of the laboratory experiments which have so far been conducted have been relatively brief in duration. We have had to rely more upon field data which is sometimes incomplete or poorly controlled. One big limitation is the composition of the groups typically studied. Most studies have been done with people who are recruited as volunteers who usually have no prior knowledge of each other. The trend has been toward studying groups of relative strangers, rather than previously well formed groups as would be the case with our astronaut crews. We really do not know what happens to already well formed, functioning groups when they are placed in isolated, confined settings. This topic for research looms large.

One of the things we do know about how new groups in isolation respond is that the familiarization process is incredibly accelerated. Studies of dyads (two people) in isolation indicate that people very quickly learn most of the mundane aspects about each other. Conversation then turns to more personal issues. The rapid revelation of such information can lead to resentments among the individuals. Also, at some point you simply run out of things to talk about. It is only a matter of time until you know so much about the other person that you cannot stand to hear the same old joke or story one more time. At this point, people tend to begin to withdraw from each other. They spend much less time interacting despite the fact that this leads to more isolation and a feeling of aloneness. Social discourse is reduced and individuals appear to become more territorial. Somehow this seems to be a means of establishing one's own privacy and maintaining social distance from the other individuals. This leads us into a discussion of privacy and some of the other aspects of craft habitability.

PRIVACY

The need for privacy by individuals in isolated, confined settings is essential. However, how can we define privacy and what are the requirements to ensure this necessity? This seems to vary as the number of people increases, as the size of the habitat varies, and with the individual personalities of the crew. Considerable research has been directed at the relationship between density (the number of people per unit space) and behavior. As far as the design of space habitats and crafts we must consider the volume of space necessary. This topic I believe has been pretty thoroughly researched. However, the more ephemoral psychological quality of privacy has yet to be completely defined or quantified. Prior to Skylab, privacy was simply nonexistent. Even on Skylab privacy needs became an issue, although on this particular project television cameras and audio equipment could be shut off when necessary. This issue is to become more prominent as we consider flights of considerably longer duration and those involving crews of mixed gender and nationalities.

HABITAT DESIGN FEATURES

In looking at some of the habitability features related to privacy, there are several suggestions that we can make based on the studies that have been done with people in confined, closed areas. One of these is that in dealing with the architecture of the colony or the capsule for a long term space flight, things should be made as changeable as possible. In fact we might use furniture or decorations which have interchangeable modules. The crew should be able to rearrange their living and even their working quarters to suit their tastes and to give them maximum individual touches. Indeed, it will be important that the living and working areas in general be as distinct as possible. This will tend to maximize the use of small space to create the atmosphere of very distinct sections within the habitat. Lighting and color designs will be particularly important in this regard.

Beyond the actual layout of the capsule we will have to make provisions for crew recreation and leisure time. An exercise and activity area will be important. Considerable research has been directed toward understanding the types of activities that people in isolation prefer to engage in. Most competitive types of activities do not appear to be used much. This is probably in response to the previously mentioned idea that people in isolation seem to avoid any type of confrontation and often prefer reduced interaction. Passtimes that are more passive in nature seem to be preferred. It will be important, then, to develop more types of individual activities.

There are a number of human factors elements related to how one can effectively work in space. Fine motor movements tend to suffer more in space than do gross movement abilities. This notion should be considered in designing display and work panels, tools, instruments, and control equipment. Pictures we see of astronauts in space appear to portray them as having an easier time of walking and working than would be the case on Earth. Actually, it is more difficult because there is often no reaction to an action initiated by the astronaut. That is, an object started in motion does not necessarily behave as it would on Earth because of the lack of gravity. The astronaut actually has to work harder sometimes. This was a problem in some of the earlier flights, particularly during extravehicular activities (EVA). On several occasions the crewmembers became so exhausted during EVA that that phase either had to be terminated early or the demands reduced. This seemed to be less of a problem in later missions where equipment designs were better.

SOCIAL NORMS AND STRUCTURE

I have alluded to the problem of ostracism in isolated groups. Let me expand upon this because I think it is a particularly important problem for future space crews. I challenge you to do a little experiment on your own. The next time you are in a group of four or five people and are discussing some controversial topic where all the other people seem to be in agreement on the issue, just try taking the opposite point of view. Psychologists have found that in such situations the first thing that the other group members will probably do is to stop talking among themselves and begin directing all their discourse toward you in an effort to change your opinion. If the group cannot succeed, they will shut you out of the conversation, talking instead among themselves to reinforce their own values and beliefs. Further, if prestige or status ratings were made of each participant after the discussion, I leave it to you to figure out where you would stand. The person with the deviant attitude or behavior is alienated. Most of us can be influenced to modify our viewpoint if sufficient social pressure is brought against us. Group processes exert a great deal of pressure toward conformity. This is especially true in an isolation setting where the prospect of social ostracism can be devastating. This is one of the reasons why confrontation is usually held to a minimum. The possibility of being alienated from the group when there is nowhere to go in space has some dramatic implications. The fact that groups can exert such a powerful influence requires us to rethink how groups in space may govern themselves or how the decision-making process may proceed.

Is it possible, for example, that future crewmembers may be more susceptible to problems of "groupthink" as it has been called than we might anticipate? Also, what affect will group processes have on the role of the leader? The leader can ill afford to isolate himself and yet may have to make more than one unpopular decision. How will this be handled by groups in space? This seems to be a large area for research that will be important to our understanding of how crews in space will live and work together.

CREW SELECTION

This is such a broad and important area that it overlaps many of the other issues related to the composition and functioning of groups in space. First off, consider that what we will have to do is somehow select and compose an effective group for space travel. The important word here is *group*. This is a totally different orientation than we have taken before. In the past we have attempted to select outstanding individuals. However, an effective individual may or may not be an effective group member. Our astronauts to date have been highly independent, concrete individuals. They have not been particularly sensitive individuals in the social sphere, preferring instead to deal with interactions from an emotional distance. This personality style has been extremely appropriate for the missions we have so far flown. We must question whether this would be the most appropriate groupmember in light of more socially demanding conditions inherent in long term space flight. Perhaps instead we need to concentrate our efforts on selecting socially oriented individuals, whose interpersonal skills will benefit the crew and mission as a whole. Our emphasis, too, should be on selecting effective groups not just individuals.

Part of our difficulty falls on the selection method. We must first decide which are the appropriate selection qualities. This is no easy matter. For example, given our choice, would we be better off to select extroverted or introverted individuals for long term flights? One might first assume that an extroverted person would be advisable. You might imagine that they will be more active in establishing and maintaining a group process. Their social needs are stronger and they might be expected to do better under the demands of a rigorous, socially monotonous environment. Actually, there is some evidence to suggest that they do not do well in social isolation. Their needs for social contact seem to be more of a liability than an asset in isolated settings. Introverts, on the other hand, have less need for social contact. Therefore they usually do not add as much to the group process. Studies of simulated isolation settings have found that an entire group composed of introverted individuals do not do particularly well. It would seem that there are still many unanswered questions regarding the issue of extroversion/introversion. This is the case with many other dimensions of personality as well.

Assuming that we can establish what qualities we wish to select for, we then have the problem of how to do this. The tests which have traditionally been used are not really designed for this approach. Our traditional psychological tests have been more frequently used to select "out" individuals than to select "in." In other words, most psychological batteries emphasize identifying undesirable traits which might disqualify a candidate rather than on identifying positive qualities we would wish our candidates to possess.

The question of test design also becomes monumental as we approach the possibility of mixed-nationality crews. It is certainly not going to be possible for us to use the same tests in all countries because of the cultural bias many tests possess. Instead, we may have to rely upon each country to select those candidates they feel are best equipped for the job. We may have to consider certain cultural factors on how individuals might be selected and how these individuals might be composed to form a functioning group. We are certainly far from answering all of these questions.

CONCLUSION

What I have tried to do is to tell you about some of the important research areas related to long duration space travel.

My intention was to ask more questions than I gave you answers to in the hopes that they might spark your interest and get you thinking about some of these issues. With the time we have had there was no way we could thoroughly cover the entire area of psychosocial issues involved in space travel. I have dealt with some and hope you have a better appreciation for the wealth of research yet to be done and the host of questions which still remain. My hope is that by trying to resolve some of the issues of space we may also resolve some of our issues here on Earth. I am quite excited about this prospect and am trying to get people to notice and deal a little more realistically with the problems involved in space. I do think that the problems are solvable, and we will learn more each step of the way. I believe that research must be concentrated on those topics I have briefly covered today. There has been a long hiatus since Skylab. The recent Soviet's long term missions have provided valuable information. The Space Shuttle's imminent launch will provide an opportunity to continue space research.

From my observations, we are in a position to begin doing some relevant space research on social and group interaction processes here on Earth. Initially, such research will have to be on a relatively small scale. Eventually, we can consider a long duration space mission of some sort, perhaps even a space colony.

ROBERT HAMMEL

Systems Development
TRW Systems and Energy

Mr. Hammel has over 25 years of experience in materials research and development and application of these developments to solving problems in man's immediate environment.

Mr. Hammel is responsible for planning and requirements for a group that is currently engaged in uranium enrichment process development; gas centrifuge enrichment process systems engineering; information systems for energy applications; nuclear waste management, isolation, and non-proliferation; and fusion-fission research.

After receiving his MS degree in Nuclear Engineering from Iowa State University in 1960, Mr. Hammel joined TRW, where he became expert in studying experimental materials and their behavior when exposed to adverse conditions, such as high temperatures, low-energy radiation, and ultra-high vacuums.

As Manager of TRW's Materials Technology Department, Mr. Hammel led a group involved in study of low-gravity materials experiments, heat pipes, and advanced alloys for eventual application to in-space programs such as the Minuteman Intercontinental Ballistic Missile, the Orbiting Geophysical Observatory, the Vela Nuclear Detection Satellite, the Apollo manned spacecraft, including the Lunar Module Descent Engine, and the Skylab.

More recently, for the NASA Marshall Spaceflight Center, Mr. Hammel led the Space Processing Payload Equipment Study, the Materials Sciences and Manufacturing in Space Payload Equipment Study and the Space Processing Applications Payload Equipment Study.

Mr. Hammel is Chairman of the AIAA Space Processing Committee and is a member of the AAS Processing of Materials in Space User/Developer Exchange Program.

4

MATERIALS PROCESSING IN SPACE

ROBERT HAMMEL‡

I welcome the opportunity to share some of my thoughts on space activities that I have had some concern with over the past several years. Things are becoming more focused with the advent of the Shuttle and its flight opportunities. This applies especially to the potentials that space may offer to the applications area of manufacturing products in space. I want to make one disclaimer right from the beginning: I am not going to talk about mining the Moon and asteroids or making large structures for the purpose of colonization. Rather, I am going to talk about some things that relate to Earth problems that we are interested in solving in the coming years.

EARTHLY PAYOFF TOMORROW

Figure 1 summarizes what NASA and Congress hope to pro-

‡The remarks by Mr. Hammel are his personal opinions and do not represent those of TRW or NASA.

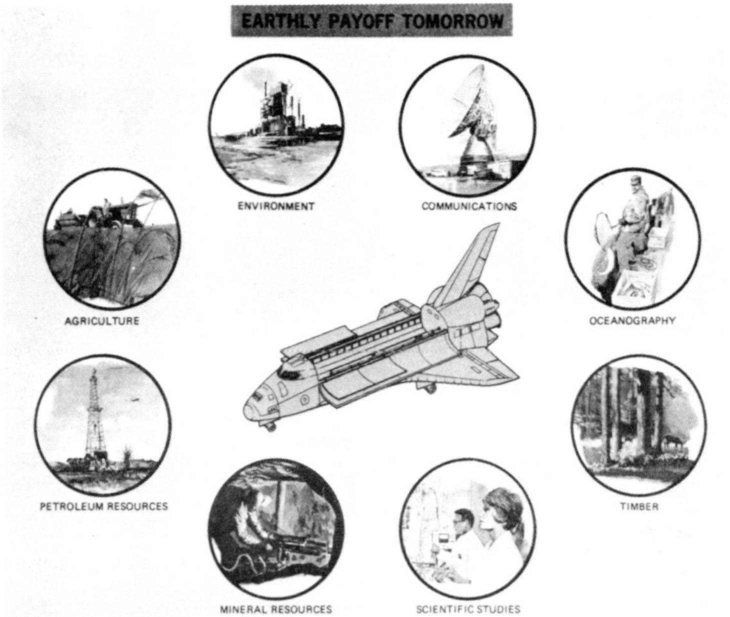

Figure 1

vide in the next ten to twenty years in applications fallout, beginning with the use of the Shuttle Transportation System. One broad area includes agriculture, environment, oceanography, petroleum resources, timber and mineral resources, all of which fall into the arena of Earth observations. We are doing things of that type now. You may be familiar with the Landsat Program where photographs are taken of vast regions of the Earth from space. Landsat, however, is not a commercial venture for no one is engaged in Landsat as a private enterprise. The area of communications, however, is in the private sector and is making a profit. Last year they sold some nine and a half billion dollars of related space services. Things have just gotten started there though, and should continue to expand into a much more important commercial venture as our desire and efficiency for data processing expands. We may someday see the wrist communicator.

Past space missions have dealt with four areas: exploration

of the Moon and the planets; measurement and observation of Earth from space; the performance of a series of military missions; and other scientific efforts. While these will continue, expansion in the applications rather than science or exploration motives is expected.

MATERIALS PROCESSING IN THE SPACE PROGRAM

The present Materials Processing in Space Program (MPS) fits under the category of science applications. It has been NASA who, over the last few years, has brought the focus to this area of endeavor. Often, however, the area of Materials Processing in Space has been misunderstood. It is my intention today to clarify some of the priorities and near-term realities of processing in space.

The objectives of the MPS program have been developed in conjunction with NASA management, Congress and the scientific and industrial community. Figure 2 lists four broad goals. They start with the fundamental objective of understanding what space offers in the scientific arena when processing materials under low gravity. This phase is then followed into follow-on demonstrations to determine potential values. Here the invention, the piece of scientific data or observation, is tested to see if a product or novel structure might be related to the application of the resulting material.

Next it is necessary to establish mechanisms that stimulate, encourage, and attract industry and the commercial entrepreneur to enter this arena. Moreover, the bankers must be convinced they should support this type of venture risk. There are tremendous governmental, management, social, and political implications in bridging this transition.

If the first three steps are successful, then the final objective is a realization that the majority of the applications activity will proceed as privately funded ventures.

These objectives will not be achieved immediately. At this

MATERIALS PROCESSING IN SPACE PROGRAM GOALS

- THE NASA PROGRAM GOALS BROADLY STATED REFLECT AN APPLICATIONS EMPHASIS BUILDING FROM BASIC INVESTIGATIONS IN SPACE

  1. DEVELOP UNDERSTANDING OF FUNDAMENTAL PROCESSES AND PROPERTIES OF MATERIALS THROUGH SCIENTIFIC UTILIZATION OF SPACE
  2. DEMONSTRATE THE VALUE OF SPACE FOR MATERIALS WORK BY DEVELOPING PRODUCTS AND/OR MATERIALS
  3. INITIATE USER-SPONSORED UTILIZATION OF SPACE FOR RESEARCH IN MATERIALS SCIENCE AND TECHNOLOGY
  4. REALIZE PRIVATELY-FUNDED ORBITAL MANUFACTURING OPERATIONS

- NASA DOES NOT INTEND TO GET INTO THE MATERIALS BUSINESS. THEY ARE SPONSORING THE NECESSARY START-UP ACTIVITIES IN THE PRIVATE SECTOR AND WILL PROVIDE SPACE FLIGHT CAPABILITIES TO THE USER COMMUNITY

Figure 2

point, I would like to expose you to my speculations about how long and what forms it might take.

BASIC TYPES OF PROCESSING

First, we need to consider the material's processes that are influenced by going into space in order to work with them in low gravity. For example, in space, a container, which is solid will float in its entirety. Not much can happen to the material making up this container in low gravity as long as it remains solid. However, a liquid that is in the container may be profoundly influenced because in the fluid state the interatomic forces are much weaker. Thus, we will be working with fluids or melted solids in space.

In space, the reduction of convective and other buoyancy factors that occur on Earth can profoundly change what happens in several processes. They generally fall into five broad categories, (figure 3):

1. Crystal Growth: In the process of simple crystal growth, you start with an unstructured array of atoms and produce a single crystal structure by causing the atoms to fall regularly into place.

BASIC TYPES OF MATERIALS PROCESSING

1. CRYSTAL GROWTH
2. PURIFICATION/SEPARATION
3. MIXING
4. SOLIDIFICATION
5. PROCESSING IN FLUIDS

Figure 3

2. Purification-Separation: Here you try to remove one material from another to achieve a greater and greater uniformity of chemical composition.
3. Mixing: On Earth there are some things we can't mix very well because forces, due to density differences, restrict mixing, causing sedimentation, layering, etc.
4. Solidification: In the process of solidification (the reverse of melting) where the material goes from a liquid to a solid, a solid-liquid interface exists at the growth surface. The ability to control the rate things go on at that interface can be profoundly affected in the presence of low gravity.
5. Processing Fluids: We have to examine closely and understand thoroughly the whole question of processing things in a fluid state. Further, we will have to develop techniques to carry out the first four processes.

By going into low gravity, we remove the conventional driving forces due to buoyancy, most notably sedimentation and convection that we always experience in our Earth-bound processes. On Earth, when you boil a pot of coffee and grounds are in it, the denser grounds go to the bottom. Bubbles rise. There is a thermal stirring action which always exists whether it be in the atmosphere, in this room, in the coffee pot or in the sea. These are all effects driven by the presence of gravity.

Now, when you go into space the bubbles don't go up. They stay in place. They may move for other reasons, but not because of the presence of gravity. Again, you can distribute particles of different density in mixtures and they stay in place. The process of the hot-cold mixing or stirring does not occur due to thermal driven convection and things do not fall to the bottom.

What happens in space is that any object that you take into orbit, has as part of its flight mechanics, essentially a continuous free fall condition. Thus, in space, an object that is not tied down will appear to float in front of your eyes. Here the materials processing scientist has another opportunity. We can take a "lump" of material: metal, glass, mixtures, etc., and free float it, controlling its position without physically putting it in a container or touching it with rods, wires or other fixtures while processing it.

How is it possible to control the position and move something about in this free-floating situation? A metal being conductive can be held in an energy well of a specially designed electro-magnetic field. If we have a material that will not conduct (insulators), and still want to work in a containerless mode with something like a glass, we can use acoustical sound waves. In other words, we can create a minimum "energy well" with an arrangement of sound drivers like the speakers on a stereo system. By properly tuning them, the object will move to a certain spot and stay there. In fact, we can also make it spin or pop up or down.

Thus, the two major features that we are dealing with in low gravity literally relate to reducing the affects of density differences such as thermally driven convections and sedimentation (either in a contained liquid that is being melted or a free floating one) and the ability to work with materials in a free floating state. In the free floating state, you eliminate undesired reactions between the material being processed and the container. Temperature then becomes the limiting case in purification or in the ability to form a number of reactive materials rather than container reactions. This will permit us to get better control of the purity and structure for materials we cannot satisfactorily handle today.

Process conditions may also be turned around to either very precisely control the heat inputs or the heat outputs under boundary conditions set up in the experimental apparatus. We do this in varying degrees of success on the Earth right now. But many of these processes in the presence of gravity, proceed rapidly through key phases with limited rate control. We believe that in space, techniques and understanding will be developed so we can control critical steps over process times orders of magnitude longer than those possible on Earth. This may enable us to precisely tailor the resulting material structures and chemical compositions.

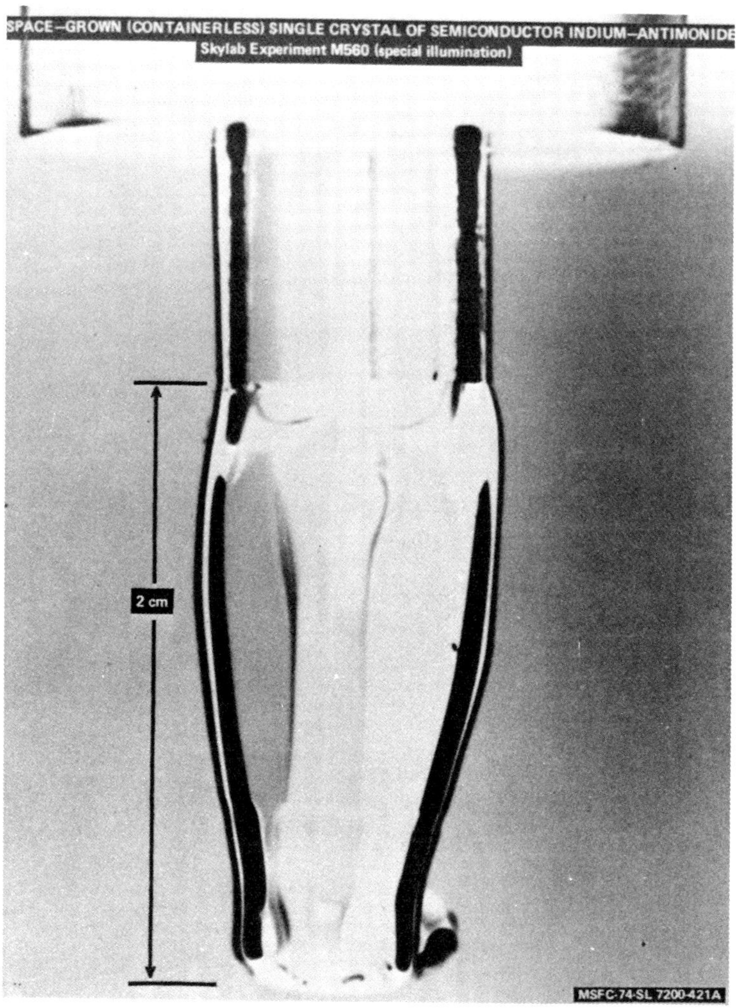

Figure 4

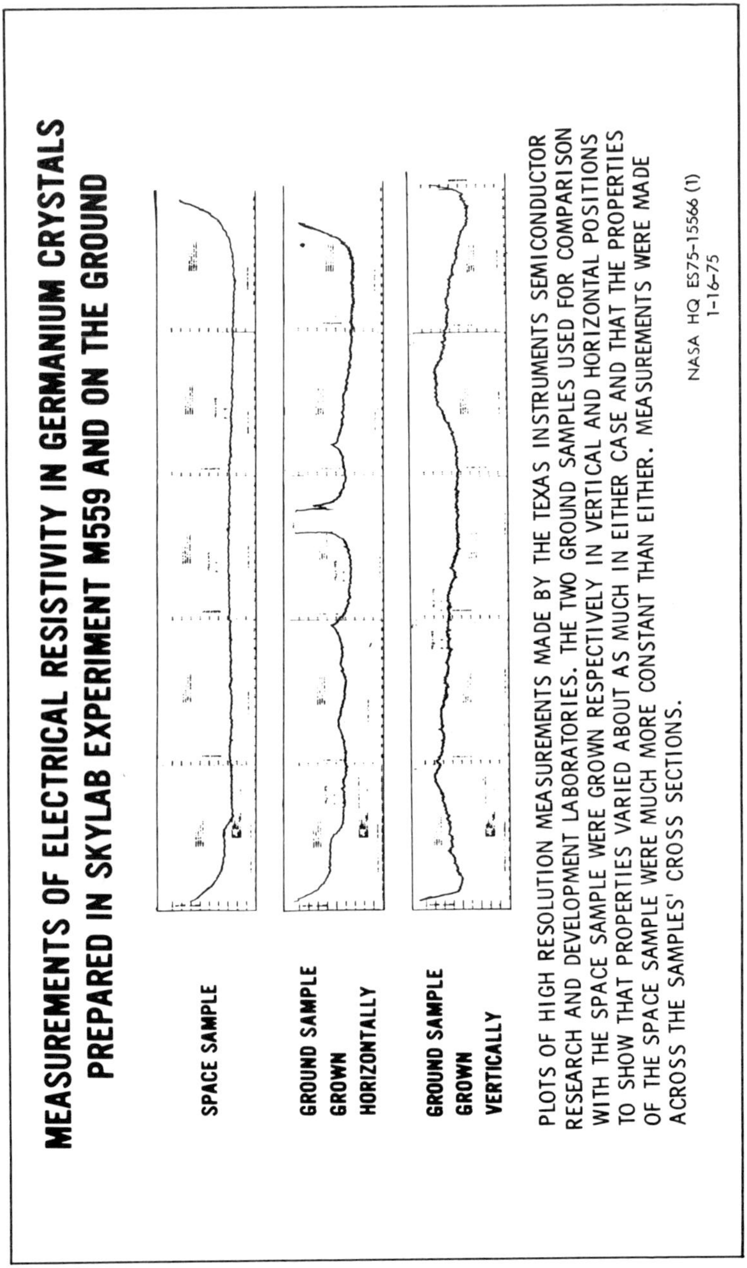
MEASUREMENTS OF ELECTRICAL RESISTIVITY IN GERMANIUM CRYSTALS
PREPARED IN SKYLAB EXPERIMENT M559 AND ON THE GROUND
SPACE SAMPLE
GROUND SAMPLE GROWN HORIZONTALLY
GROUND SAMPLE GROWN VERTICALLY
PLOTS OF HIGH RESOLUTION MEASUREMENTS MADE BY THE TEXAS INSTRUMENTS SEMICONDUCTOR RESEARCH AND DEVELOPMENT LABORATORIES. THE TWO GROUND SAMPLES USED FOR COMPARISON WITH THE SPACE SAMPLE WERE GROWN RESPECTIVELY IN VERTICAL AND HORIZONTAL POSITIONS TO SHOW THAT PROPERTIES VARIED ABOUT AS MUCH IN EITHER CASE AND THAT THE PROPERTIES OF THE SPACE SAMPLE WERE MUCH MORE CONSTANT THAN EITHER. MEASUREMENTS WERE MADE ACROSS THE SAMPLES' CROSS SECTIONS.
NASA HQ ES75-15566 (1)
1-16-75

Figure 5

Crystals are one important example of this desire. Crystals prepared on Earth were taken into space on Skylab and remelted (see figure 4 for example). When we grow or produce certain single crystals on Earth, they often look like a cylinder. Under careful examination, one can see frozen bands or layers in the solid that formed as growth proceeded.

The perfection of the crystals processed in Skylab were amazing because they were free from numerous defects contained in comparison parts which were processed on Earth. This lack of defects is due to how uniformly the individual atoms stacked up. In figure 5 we see results comparing the resistivity of germanium crystals prepared on Earth and in space by the same process and then slicing each individual crystal along its long axis for characterization. The space-grown crystal, other than end effects, shows very little variation in resistance along the crystal.

The problems of varying properties we have with Earth-grown crystals is because of spacial differences both in distribution of the compounds and in microscopic growth rate fluctuations at the growth interface produced during growth. The data for both the horizontally or vertically Earth-grown crystals of figure 5 illustrate such variations in uniformity. Growth banding in Earth-grown crystals results from the fact that as the crystal is growing, for a very small instant the liquid-solid interface is super heated because it is trying to take a liquid and form a solid. Then it super cools, so there is a small thermal racheting going on. Each of these represents a little localized convective instability that progresses stepwise as the crystal grows. In space this process disappears. Instead of having convection disturbed growth we are in a diffusion controlled regime. Here the heat transfer and the rate at which the materials come to the growing interface are very stable and precisely controlled by predicable laws of physics.

PRIOR EXPERIMENTAL WORK

The Materials Processing in Space Program has been around in several experimental forms since 1971. As we note in figure 6, the first experiments in the low gravity behavior of materials were conducted on the Apollo 14 mission. During the mid-course per-

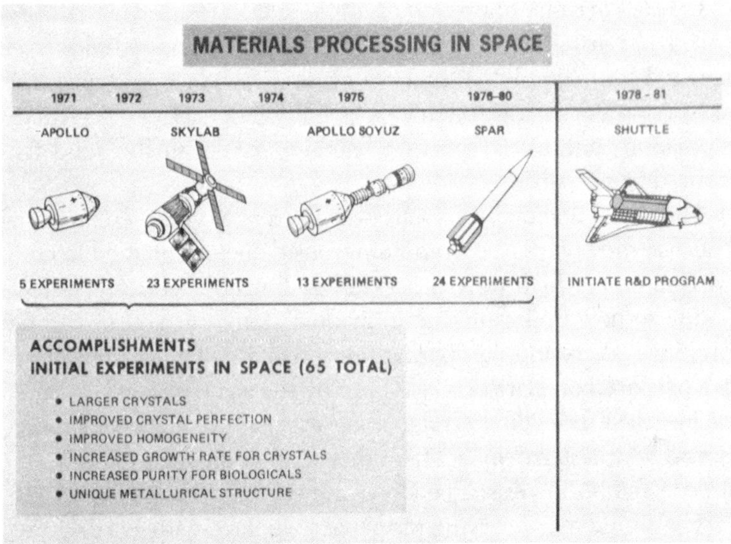

Figure 6

iods while they were coasting to the Moon or coming back, the astronauts had time to do some simple demonstrations. These included experiments in solidification, heat transfer, and simple biological separations. A more comprehensive series of experiments was conducted on all three of the Skylab missions. Other experiments continued using sounding rockets to get a few minutes of zero gravity in a ballistic trajectory, a drop tower producing a few seconds of low gravity free fall, and in aircraft flights put into a special path producing a twenty to thirty second reduced gravity period.

Some 65 total experiments up to this time have simply whetted our technical curiosity. It has shown us much about what we need to learn and do in order to go into space to work with materials. There have been advances in our experimental equipment, in the experiment theories, and in learning how to prepare experiments. Many more are yet to come. What we have done, has led us to where we are ready to initiate a new era where we hope to make a much more serious penetration into the total field of low gravity processing of materials.

THE SHUTTLE

The next step in the process of advances is by means of the Shuttle. What is the significance of the Shuttle and where does it take us in terms of a scenario? The Shuttle offers a vehicle (Spacelab) which allows us to take a pressurized manned laboratory environment into space for a short time to experiment in various science

Figure 7

Figure 7

and applications disciplines. The initial MPS planning by NASA intends to implement by the early 80's two types of mission capabilities. In Spacelab racks, figure 7, groupings of apparatus that several investigators will use on a given flight may satisfy such MPS research interests as solution crystal growth and bio-processing. The Shuttle will also carry things on an unpressurized pallet in the Shuttle cargo area, so that when the doors are open, the pallet payload is directly exposed to space. An early MPS payload is also planned for this approach.

A feature of the Shuttle you don't often hear about is the available power. When Shuttle flies, it will provide on average up to seven kilowatts of power for whatever is in the Shuttle cargo bay while on orbit. Seven kilowatts may sound like a great deal, but few people realize how much power is available to a home electric oven. About forty kilowatts. A typical home is wired for 50 to 100 kilowatts of power and can be drawn on anytime. Usually, we can run everything at the same time. In space, the Shuttle, while bigger than the average house, is very power constrained. It takes about four kilowatts of the seven to run the environmental control system — the air conditioning and the lights, etc. of the Spacelab — leaving three for the experiments.

Materials processing which requires high temperature activities for melting metals and performing certain crystal growths will build an automated apparatus to be flown on a pallet as shown in figure 8. It will receive somewhat more experiment power than

Figure 8

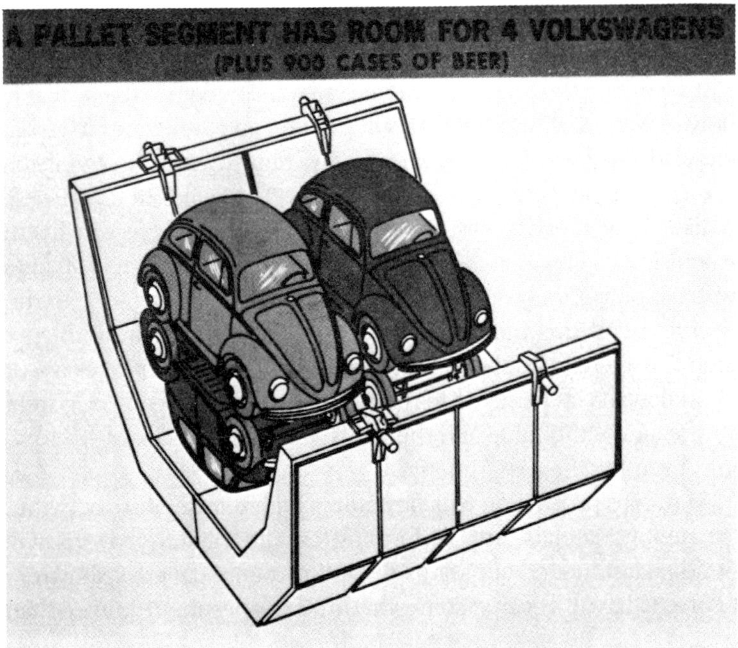

Figure 9

the Spacelab mode but poses another constraint to the apparatus design. The crewman on the flight deck will push buttons and tweak knobs to cause things to happen on the pallet equipment, but there will be no direct hands-on capability. The crewman will remotely observe certain parts of the experiments. In order to give an idea of the size of a pallet, we did a simple calculation which showed that you can, if packed right, carry four Volkswagen Beetles and 900 cases of beer. (figure 9).

Figure 10 shows the TRW built Viking Life Detecting Experiment which landed on Mars. It consisted of a miniature laboratory that did many things. Shown in the background is exactly the same capability built with conventional laboratory apparatus found in a high quality Earth-bound laboratory. It does all the same science that the small package does, yet the Viking unit is a little over a foot on a side. The difference in cost was tremendous to go to this reduced size and get the reliability necessary to assure it was reliably

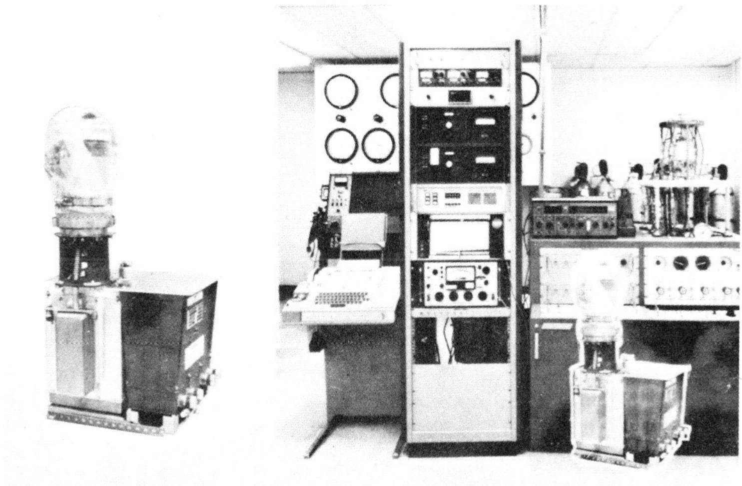

VIKING BIOLOGICAL EXPERIMENT DESIGNED FOR VIKING

VIKING BIOLOGICAL EXPERIMENT DESIGNED FOR SPACELAB

Figure 10

going to work when it arrived at Mars.

The Spacelab and the Shuttle Transportation System allow us to depart from some size and reliability constraints (not safety) in developing payloads. That is an important factor. It is a feature that says we can afford to make a mistake in designing the initial concept because we can bring it back to fix or alter it. We don't have to do all the miniaturization, which is a very costly process. Instead the developments for MPS will be in techniques and hardware that will work in low gravity to do the processing we want.

A Multi-Purpose Fluid Phenomenon Facility is one of several units TRW has been put under contract to develop for the initial Shuttle Spacelab flights. It will be used to measure and observe fluid phenomena such as crystal growth, bubble dynamics, and particles in microgravity (figure 11). The other Materials Processing payload being developed to fly on a pallet will examine high temperature materials of interest to electronic devices, optical components, glass products, ceramic products and special metals (figure 8).

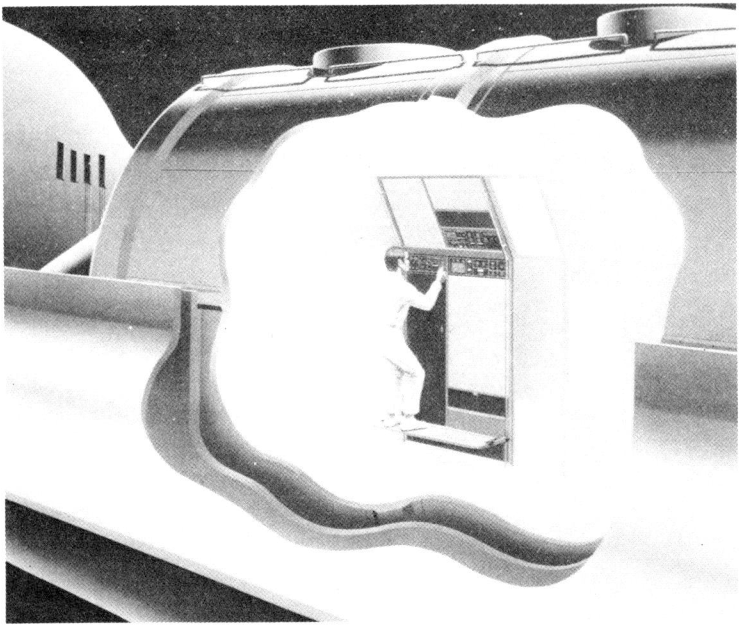

Figure 11

WHAT HAS TO HAPPEN

Does this mean "we are about to have a factory in space?" or "We are going to manufacture in space?" The answer is not simple (figures 12 and 13). Certainly not immediately. First, we must obtain further indication that we really have a unique environment for processing materials in space. We certainly have some strong clues from the results of Apollo and Skylab, but we are still at a pioneering stage. We need to understand the unique nature of space processing and how to control that uniqueness in the experiments coming up. We must get accurate scientific results which will support our looking into the degree gravity is a factor in the manufacture of possible products. Such research alone is not going to tell us what the products of future space manufacturing can really be, because a complex set of marketing and economic arguments must also be addressed. Good science is only one factor to be considered.

Progress in materials processing has a need for a large experimental content. Initial experiments were based on the

INDUSTRIAL SPACE PROCESSING

IN THE BEGINNING:

1. OBTAIN SUFFICIENT INDICATION THAT SPACE PROVIDES A UNIQUE ENVIRONMENT FOR PROCESSING.
2. DETERMINE WHETHER GRAVITY IS THE LIMITING FACTOR WHICH PRECLUDES MANUFACTURE OF PRODUCTS ON EARTH.
3. DEVELOP A BODY OF THEORETICAL KNOWLEDGE UNDERLYING THE SIGNIFICANT ADVANTAGES IN SPACE.
4. OBTAIN EXPERIMENTAL EVIDENCE INDICATING THAT THESE THEORETICAL ADVANTAGES CAN BE REALIZED.
5. PROCEED WITH THE DEVELOPMENT OF SPACE HARDWARE SUITABLE FOR THE INVESTIGATION OF BASIC PHENOMENA BASED ON CURRENT KNOWLEDGE AND ASSUMPTIONS.

Figure 12

INDUSTRIAL SPACE PROCESSING

ALSO REQUIRED:

1. DEVELOP THE TECHNOLOGY WHICH PERMITS RECOGNITION OF THE PRODUCT FORM.
2. CAUSE THE "USER COMMUNITY" TO SERIOUSLY CONSIDER THE RESOURCE OF SPACE AS AN ADDITIONAL RESEARCH TOOL.
3. IDENTIFY POTENTIAL TARGETS OF OPPORTUNITY AMONG ECONOMICALLY SIGNIFICANT PRODUCTS FOR EARLY COMMERCIALIZATION.
4. CONDUCT AN ADEQUATE NUMBER OF MEANINGFUL DEMONSTRATIONS TO CONVINCE INDUSTRY THAT PROFIT POTENTIAL EXISTS.
5. DISSEMINATE THIS INFORMATION IN SUCH A MANNER THAT IT BECOMES WIDELY AVAILABLE TO POTENTIAL USERS, WHILE STILL PROTECTING THE PROPRIETARY RIGHTS OF THE INITIAL ENTREPRENEURS.

Figure 13

somewhat Edisonian rationale: We know there is something of interest. Let's go do something in space. We'll make an exact duplicate of the experiment on the ground, but it won't be in low gravity. We'll cut up the resulting specimens and look at the data, and if they are different, we will try to figure out why they are different.

That is not the way you do good science. I don't care whether it is in sociology or physics, you have to have used better planned experiments, controls and supporting rationale to allow unambiguous interpretation of results. In the next step, space materials science has to mature by expanding the appropriate theoretical models. There is a lot of current theory and equations that say we understand much about materials processing, but we have to translate this to terms that can be verified under low gravity. Then we have to confirm what can be practically realized by process experiments. The latter will require technology and some engineering developments to provide the apparatus. Lastly, a sustaining commitment is necessary to gather the information; to take apparatus into space and to collect and build that much broader body of information. That's the necessary applications foundation — the front end work yet to be completed.

Figure 14 represents a popular chart I have used in several

POTENTIAL SPACE PRODUCTS

ELECTRONIC PRODUCTS	OPTICAL PRODUCTS	BIOLOGICAL PRODUCTS	STRUCTURAL USES
SEMICONDUCTORS	NARROW BAND FILTERS	HIGH-PURITY BIOLOGICALS FOR USE IN MAKING VACCINES	BETTER TURBINE BLADE MATERIALS
COMPOUND SEMICONDUCTORS	LARGE DIAMETER CRYSTALS	HUMAN CELL PURIFICATION	STRONGER MATERIALS WITH LOW EROSION RATES FOR SPECIAL APPLICATION, INCLUDING PINS FOR BONE SETTING
INTEGRATED CIRCUIT CHIPS	UNIFORMLY DOPED CRYSTALS	ENZYME ISOLATION (PRODUCTION OF UROKINASE)	LOW-FAILURE-RATE MATERIALS
MAGNETIC SWITCHES	SELECTED WAVELENGTH OPTICALS	PROTEIN PURIFICATION AND PRODUCTION	LONGER LIFETIME MATERIALS
RELAYS	SUPPORTING OPTICS FOR LASER SYSTEMS	BIOLOGICAL FACTORS IDENTIFICATION	IMPROVED LUBRICANTS
MAGNETIC DETECTORS	HIGH INDEX OF REFRACTION GLASSES	HORMONES PRODUCTION (ERYTHROPOIETIN)	HIGH STRENGTH COMPOSITES
ULTRASONIC AND OPTICAL FREQUENCY FILTERS	IR-TRANSMITTING GLASSES	IMMUNOGLOBULIN PRODUCTION	
SUPERCONDUCTORS	IMPROVED FIBER OPTICS	DIFFERENTIATION, ISOLATION, AND CULTURE OF WHITE BLOOD CELLS (LYMPHOCYTES)	
HIGH POWER RECTIFIERS	CERAMIC WAVEGUIDES		

Figure 14

publications. The title says, "Potential Space Products." Somehow people forget the first word of the title. What they say is, "Look at all those products!" There are a wide variety and number of things listed. First we find that there are application categories with potential high value added typifying the thrust of present day anticipated material breakthroughs. These generally fall into electronic products, optical products, and biological products. The title could just as well read, "Interesting Scientific Categories of Research" . . . and under that title I could list a number of things happening that *may* become important to us in terms of our future technological advances.

As in the past, we believe the future holds many new materials that will qualify for these or other projected product uses, but they will only be identified once necessary research has been done. Today there is no one who has sufficient knowledge or information to support a claim any one of the examples on this chart will become a product of space-making it through all the practical steps yet to occur.

From R&D to product form is a tough game. It's currently tough here on Earth, and when you go into space it gets even tougher. Let us assume that in space the materials science and technology go well; there are other constraints. To substantiate viability of a space manufactured product requires production machinery and new types of data on scaled up work activities in space. A "patent" and investigation in the laboratory "does not a product make."

Another aspect of future space product development requires the identification of a user community. I am a user; you are a user community. The user community is not just people in the business of selling space hardware; or just scientists who have a scientific curiosity; or just project officers who have a military mission. A user community for MPS must also include those who have not routinely thought of space as something that they can achieve benefit from or use.

We need to identify potential participants from industrial companies such as Monsanto or 3M who would be interested in the possibility for commercial products and would be willing to go through the process of testing the possibilities for their businesses. If they can argue they can make something better, then what will it cost? That requires a time frame of work in space and a series of government, scientific, and entrepreneural communications that have yet to be created. There must be consideration for the legal aspects, patent rights, and above all, sensitivity and awareness of the differing interests of the various contributors throughout a complex progression.

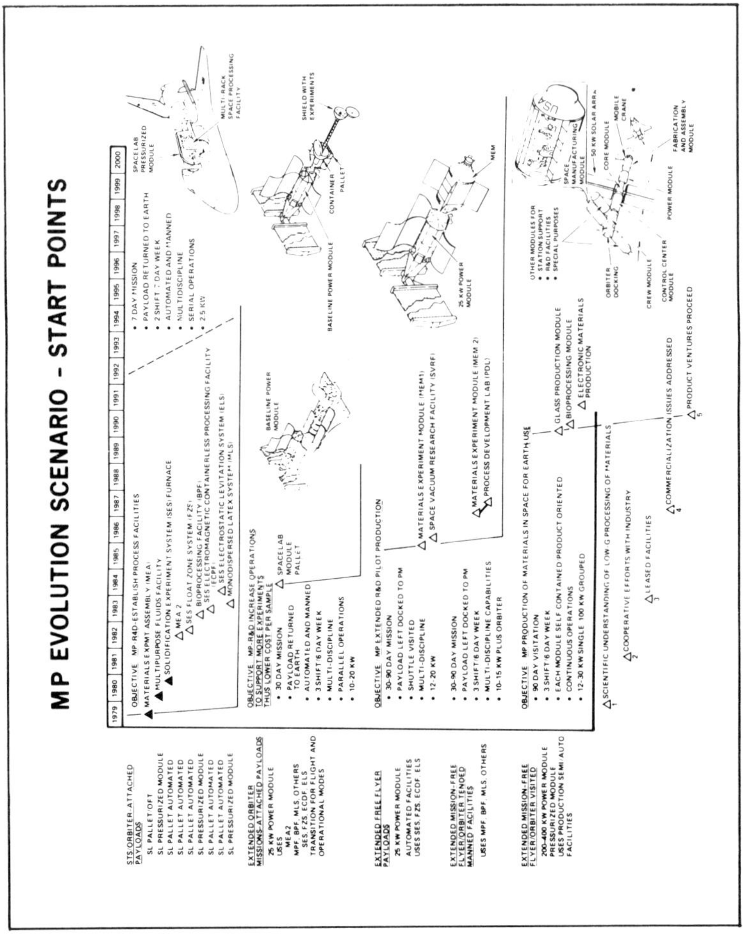

Figure 15

Figure 15 typifies a potential step-like capability development for space materials processing. The time bar spans twenty years til the year 2000. The left hand side contains a series of processing facilities being considered for development. A series of progressions will be important. The Shuttle Sortie flights starting out in the early 1980's will definitely be flown. These are seven day missions and the payload is returned to Earth. These early missions will provide limited power and operating times severely restricting the MPS technical program. NASA has recognized this flight constraint is going to be difficult to deal with including the high associated launch costs to use research equipment for only a very short period of orbital time. Current studies are considering a space unit increasing the available power up to 25 kilowatts. The Shuttle would dock with a space depot and operate payloads while docked or leave them behind to operate over extended times in space until returning.

Another suggested concept would use part of the Shuttle payload for an additional solar power unit. This would give the Shuttle a 30 day mission capability providing more than an order of magnitude increase over the seven day mission. Such advances would provide much more space work. Energy and time are needed to do the necessary research and development capable of supporting evaluations of product feasibility and the cost of doing prototype production in space.

The final facility shown in this evolutionary scenario in the "MEM", Materials Experimental Module, which automated or later as a manned version, would be docked and left as a mini-station with servicing from Earth for resupply and crew change.

In this evolutionary process, we start with the R&D (Research and Development) shorter missions but quickly, even for continued R&D, expand them significantly. If we are seriously going to address work in space issues and establish a large body of technical data, we force ourselves to consider the need for and types of additional flight capabilities. Each step in the process must answer critical questions of a scientific and then product-by-product basis. The fact that the space machinery begins to grow up, of course, leads to a broader and broader body of participation within the total community described earlier.

If some form of early space factory concept comes about as an Earth orbiting facility in the 2000's or so, it will have to account for just about all the labor relations, management and other issues that we deal with on Earth. We will have problems to deal with in human motivation because these flights are not going to be like the original exploratory space missions. Putting people in an initial space work environment, which may be viewed as confining, may cause problems.

TRW plans to continue to work closely with NASA in helping to develop our physical capability. Over the last several years we have been working with several other industrial companies such as the 3M Company, Beckman, US Steel, Owens-Illinois, and others on what space portends. If you remember what happened in the semi-conductor industry, semi-conductor materials were invented, Nobel Prizes given, and people started making products. The companies big in that area today did not exist at the time the invention occurred — Texas Instruments, notably. The companies that were big in the vacuum tube industry, said, "I've got the business, why worry about all that new stuff." They are pretty much out of that kind of business these days. It was the little companies who seriously picked up on the new technology; who had that entrepeneural spirit — some of them high rollers — took whatever "venture capital" they could muster and rode the crest, starting out as Chrysler or General Motors did in 1919 in automobiles.

It is not always easy to motivate the TRW's, Owens-Illinois or the U.S. Steels because they are large and conservative. They are already successful and generally their management are not sitting there asking, "Where is our next big chance?" But here is one very important difference between the prior development of semi-conductor electronics, and materials processed in space — the cost of money to start something in space. The one thing that really could hamper an industrial space processing program could be financing the risks. The role of the industrial participant, small or large, is yet to be determined.

IN CLOSING

My fundamental message was graphically captured by Alex MacLean of UPI (see Figure 16), "As our business goes up, we must keep both feet on the ground." In a pragmatic sense there are exciting things to do. We will do some research and we will take the opportunity to go up and explore, but if we are to mature in the applications area, we must not forget some very practical things are going to have to be achieved.

Figure 16

G. M. HANLEY

Chief Project Engineer, Advanced Programs and Technology, Rockwell International Space Division

Mr. Hanley, presently with Rockwell International, is Chief Project Engineer of Advanced Programs and Technology in the Space Division. Prior to that, he was Manager of the Rockwell Satellite Power System Study Contract with NASA. He has also been Manager of Technology Programs and Energy Systems Engineering, responsible for all the technological effort related to Energy Programs for Space Development. He has also managed Advanced Programs such as Orbit-to-Orbit Transfer Stages (Space Tug, Orbit-to-Orbit Stage, Solar Electrical Propulsion Stage), and the One Man Lunar Flying Vehicle. Mr. Hanley was Project Manager of Re-Entry Systems responsible for the conception and preliminary development of Advanced Manned and Un-Manned Re-entry Systems.

Mr. Hanley received his B.A. and M.A. in Aeronautical Engineering from the University of Minnesota, and has published numerous papers on Energy, Space, and Re-entry Systems.

5

SPACE SHUTTLE AND SOLAR POWER SATELLITE SYSTEMS

G. M. HANLEY

Things have changed considerably since we started the space program by putting men up into orbit in the Mercury capsule. We have become much more "practical" — restrained more by what people need rather than by what we would like to do. Space Industrialization — the potential utilization of space — is the talk of the future, and a major component of that future is the solar power satellite system, and the vehicle that will make it possible, the Space Shuttle.

THE SPACE SHUTTLE

The most important fact about the Shuttle is that it makes it possible to get men and materials up into orbit in a routine manner, taking payloads of up to 65,000 pounds each time.

In its launch configuration, the Shuttle itself, with its 15' by 60' payload bay is flanked by two solid rocket booster engines, and sits on the main fuel tank with its liquid oxygen and liquid

Figure 1

hydrogen. At launch, all three rocket systems fire — the two solid rocket boosters and the three Shuttle engines located aft in the Orbiter. At intermediate velocity the two solid rocket boosters are "staged off" and they fall into the Atlantic Ocean where they are to be picked up by barges and returned to Cape Kennedy where they will be refurbished to be used again. The three engines on the Orbiter continue to burn until 50 miles altitude is reached and the liquid propellants in the main tank have been completely used. At that time the main tank is separated and drops back to be burned up as it plunges into the atmosphere.

At about 100 to 300 nautical miles, the Orbital Maneuvering System inserts the Shuttle into its orbit and work begins. Once the payload is deployed, the Shuttle de-orbits and re-enters the atmosphere using the tiles on the lower surface to radiate the re-entry heat away, and lands again. With the guidance and navigation systems we now have, even though the Shuttle uses a "dead stick" landing, chances of a mistake are pretty small. Once back, the payload is unloaded, it is refurbished and we start again.

The Shuttle system makes it possible to do a lot of things we could not do before. The Shuttle can bring up larger payloads, and at much lower cost than ever before, bring payloads back down from orbit for repair or study, and provide a capability of manned interaction with payloads on orbit.

THE SOLAR POWER SATELLITE SYSTEM

CONCEPT OVERVIEW

Solar power satellites use solar cells to convert solar energy on orbit into electrical energy which is transmitted down to the ground using radio frequency waves which are received by a ground antenna and turned into direct current electrical energy.

The satellite is in geosynchronous orbit which means the spacecraft is stationary relative to a point on the Earth, keeping it directly over the same location all the time. At this altitude sunlight is present approximately 98% of the time — the 2% occurring when the Earth comes between the Sun and the satellite.

A typical satellite is roughly 10 miles long and about 2½ miles wide — about the size of Manhattan Island. Obviously, it doesn't come up in the Shuttle. It is assembled on orbit, with the materials brought up from Earth. If 60 of these projected satellites are constructed according to plan, they will produce 15% to 20% of the electrical energy needs for the year 2030. Each satellite produces 5 gigawatts of power (that's 5 billion watts) so 1½ satellites would supply enough power for Los Angeles or other large cities like New York or Chicago.

Figure 2

Some people have asked about the reason for putting these solar arrays on orbit instead of on the Earth in the first place. First of all, if this structure was on the ground it would require a very large area because solar energy is so diffuse on the Earth. There are day/night cycles, clouds, and gravity. The gravity is particularly detrimental for Earth based structures because they could weigh 100 times more than their equivalent in orbit.

Plans are to start construction on solar power satellites

around the year 2000, and to have about 60 around the Earth by 2030, each designed to last for 30 years in orbit. Currently both Rockwell International and Boeing are contracted to do studies through the National Aeronautics and Space Administration (NASA) for the Department of Energy in terms of feasibility and cost to the consumer relative to other potential energy systems. Over the last 2½ years we have investigated a number of systems, finally focusing on a solar cell design.

DISCARDED SPACE SYSTEMS

Nuclear

After looking at the possibility of putting nuclear reactors up in orbit, we found a potential safety hazard for space habitation. If there was a "burn-through" of a reactor in a 24 hour orbit, that orbit could be contaminated against any future use because the radiation would be diffused over very large distances

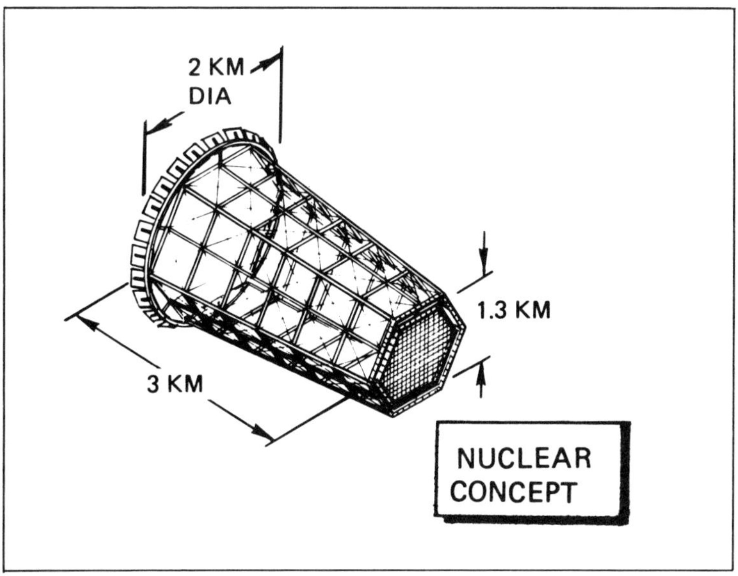

Figure 3

and could not be contained to a specific region as it could be on Earth. In terms of producing energy on orbit, nuclear reactors would work the same way they work on the ground, using nuclear energy to produce heat, and using turbo machinery to generate mechanical energy which is changed into radio frequencies and sent down to Earth as microwaves. There does not seem to be any advantage of putting fission systems into orbit by making them smaller or lighter. However, there may be compelling reasons to put fusion into orbit because of the vacuum. In addition we found greater political problems than engineering ones, so it was decided to discontinue the study of nuclear systems.

Solar Thermal Concept

Basically the solar thermal concept uses solar energy to create heat energy. This turns turbo machinery to generate electrical energy which is then sent down to Earth as microwaves.

The solar thermal system has a parabolic reflector pointed at the Sun. When the sunlight hits the reflector it is reflected up into a receiver which has a compound parabolic concentrator that throws stray light into the receiver. Boilers are located on

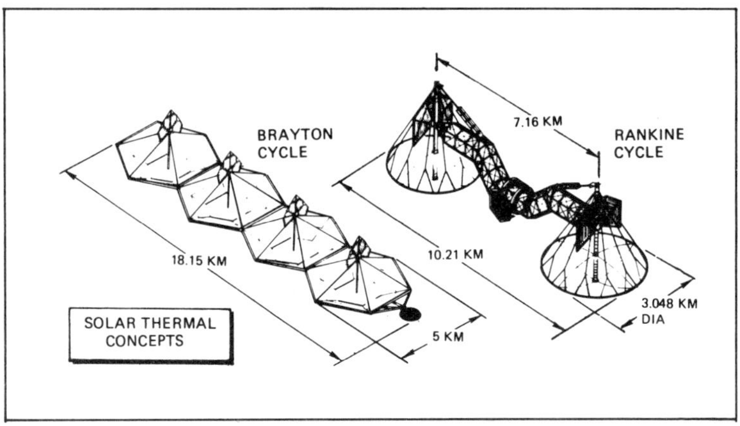

Figure 4

an absorbing surface that has a working fluid in it. Independent power modules are located on the opposite side of the absorber. The heated Cesium fluid goes through the turbo-machinery after which it heats up steam. This turns another turbine system. Eventually the steam goes up into the radiators, condenses to dissipate waste heat and is used again. The electrical power that is generated is transformed into microwaves to be sent to the ground station.

A thermal system weighs about 42 million kilograms (106 million pounds) and is 13 kilometers by 6 kilometers (8 miles by 3 miles by 3.7 miles) in size. Interestingly enough, about half of the power collection mass is in the radiator system so it turns out to be heavier and more expensive than the photovoltaic systems.

PREFERRED CONCEPT: PHOTOVOLTAIC

Up until now, silicon photovoltaic cells have been used in spacecraft, but there is a new type now in the laboratory that is very near production and will be better than silicon cells. One of these is a Gallium Aluminum Arsenide (GaA1As) cell which is especially suited for use in space. Silicon cells, when they are in orbit for a long period, decay in efficiency because of particulate radiation in about 10 years. Gallium Aluminum Arsenide cells, however, are half the weight of the thinnest silicon cell ever made and they are resistant to radiation. If you radiate them with the protons and electrons normally found in the plasma of space, their degradation is less than that of silicon cells. Furthermore, the normal operating temperature of a solar satellite using reflectors to concentrate to two suns is 125°C. The Gallium Aluminum Arsenide cells are self-annealing at this temperature (they return to their normal condition) and so they never really degrade. They are particularly suited to in-orbit use.

The Gallium Aluminum Arsenide cells are projected to be only a few microns thick, and are thus much thinner and lighter than silicon cells. This is made possible by a vapor disposition process used to deposit the film on a substrate material.

The Gallium Aluminum Arsenide comes in as a chemical vapor and is sprayed on synthetic sapphire at a temperature that keeps it from adhering to the machinery. The final solar cell is only 78 microns (3 mils) in thickness including the substrate that you attach it to and the kapton that supports the cell. The working part of the Gallium Aluminum Arsenide cell is only a few microns thick.

The Gallium Aluminum Arsenide cells are located at the base of three troughs with reflectors on either side composed of a ½ mil aluminized kapton which results in very light reflectors. The reason there are three collectors rather than one very large one is largely due to the need to keep the length short so that power distribution mass does not become exorbitant.

A satellite like this will produce 5 gigawatts of power at the utility interface. The solar energy input to the satellite is 82 gigawatts. There are numerous elements that go into this efficiency chain. Seasonal variations take off some 10%. Seasonal

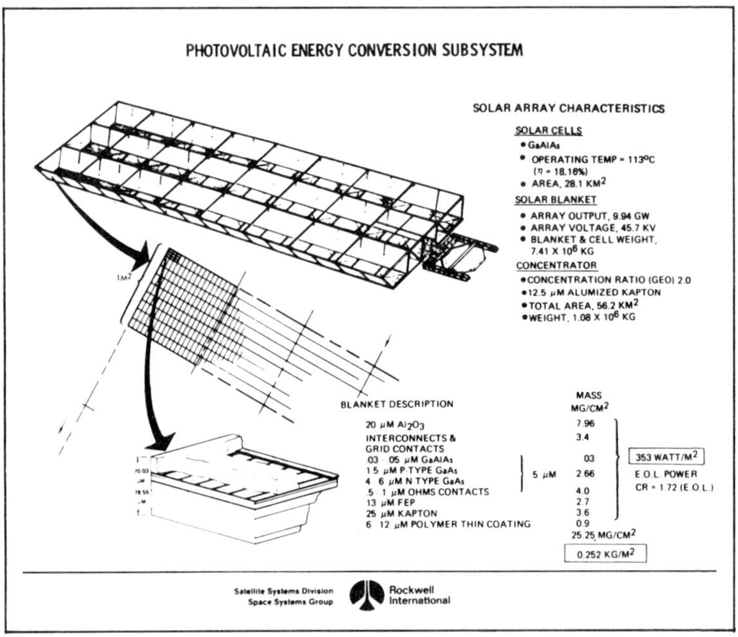

Figure 5

here does not mean summer/winter, but refers to the fact that the Sun goes north and south across the equator some 23.5° from summer to winter and back to summer again, and the required placement of the configuration results in the 10% loss of efficiency. The concentrators keep their reflectivity quite well over the projected 30 years of satellite use, giving an 86% efficiency. The solar array is about 15% efficient. Distributing the power from the arrays to the microwave system is something like 93% efficient. The microwave antenna which turns direct current energy into radio frequency is 78% efficient, and propagating that to the ground it is 83% efficient. When it is received at the antenna it is 85% efficient. Relating all of that together you get 6.08% efficiency.

The satellite antenna is one kilometer (.6 mile) across and is built that large to focus the beam down on the Earth. A small antenna would diffuse the beam. It's much like an optical problem. The bigger you make the optics, the smaller you can make the spot. Focus control is an important factor.

The weight of the satellite is 38 million kilograms (83.7 million pounds) as compared to 48 million kilograms (105.8 million pounds) for the silicon cell system. In the collector array most of the weight is in the cells themselves. The structural part of the array is only 3.8 million kilograms (8.3 million pounds) out of the total of 14 million kilograms (31 million pounds) for the whole array. Most of the weight is actually in the operating sub-systems which produce the energy. The benign environment results in an extremely lacy, light structure. The klystrons used to produce the microwave power weigh about 7 million kilograms (15 million pounds).

Figure 6 shows a cross section of a solar cell. Two bays of solar cells are put together in a series to give us the voltage that is needed to bring power into the central antenna. Each of these comes into the microwave antenna independently so that if there are problems in one of the circuits the whole satellite doesn't have to be turned off to offset the one problem circuit. When the power comes in from the solar array circuits at 45,000 volts, it is transferred to converters located on the antenna which bring it down to the voltages suitable for operation of the klystrons, where it is transformed into microwave energy.

The microwave antenna is located in the center of the satellite and is always pointed down toward the Earth receiving station. The external structure of the antenna looks like a tennis racket with lacing going across in both directions — about 1.2 kilometers (.75 mile) from end to end. Attached to each of the squares of the webbing are mechanical modules which are about 30 meters (90 feet) on a side. Each of these mechanical modules in turn is made up of a sub-array of about 10 meters (32.8 feet) to a side. The sub-arrays are composed of the power modules each of which has a klystron tube in it — with a total of 136,000 klystrons in the antenna (no small installation job). The klystrons function by taking the direct current energy and turning it into radio frequency energy which is radiated out through the slots on the front of the antenna. The antenna itself acts to radiate off the waste heat coming from the working

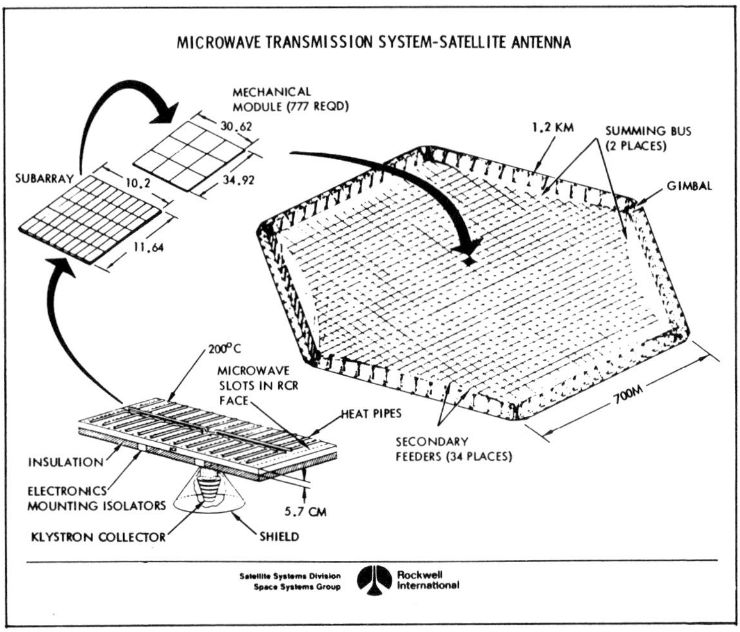

Figure 6

klystrons by spreading the heat out over the entire antenna and radiating it out into space.

How much microwave radiation actually comes off this antenna? People become concerned about microwave radiation and what they think it might do. They think about microwave ovens with the implicit dangers they have heard about them. However, the energy sent down from space is by no means comparable to the energy used in a microwave oven. At the center of the beam as it reaches the Earth there are 23 milliwatts per centimeter squared (23mw/cm^2) which is ¼ the solar constant, meaning that the microwave energy is comparable to ¼ the amount of energy you get from the Sun when you are standing outside — a very small amount. The United States standard for safe exposure outside a working microwave oven is 10 mw/cm^2. At the center of our beam, then, there is only 2.3 times this maximum standard while there is only 0.1 mw/cm^2 at the outer fenceline of the receiving station. This is much less than the U.S. standard.

The potential danger to people is not the major area of concern for the effects of microwaves however. What we are worried about is a possible interference with the ionsphere — the D and F layers. The D and F layers are used to bounce off other radio frequencies used by the military. If this layer is heated up it may create a "hole" so that these radio signals would no longer "bounce off" but would go through — a form of radio frequency interference. Sixty satellites with 60 holes could cause serious problems. But we think that by using 23mw/cm^2 we may avoid this problem. Another potential problem is the "harmonics" at the 2.45 gigahertz band that come off any system that is producing electrical energy. These may interfere with other radio frequency bands.

Some people have also asked about using alternatives to the microwave energy beams such as lasers. Current studies show that lasers are not effective enough to be competitive with the radio frequency microwaves. Lasers would have problems with clouds. At 2.45 gigahertz (2.45 billions of cycles per second), microwaves have a completely virtually open atmosphere — a window that goes through clouds, rain, and hail.

There would be some scattering, but not much, and there would be a continuous power generation with only brief interruptions about 2% of the time due to solar eclipses which occur at low peak use times (midnight).

The Rectenna

Typically, the rectenna on the ground will be 10 km (6.2 miles x 7.5 miles) by 13 km and will put out some 5 gigawatts of power. Essentially, it will look like large billboards laid side by side at a 40° angle in some 814 rows. Each panel is about 36 feet on a side. These rectenna modules receive the incoming microwave energy, much like a series of TV antennas would, with an efficiency rate of 85%. Once received, the energy is brought out of a dipole antenna and transformed to direct current energy by a diode. This power is collected in a central distribution system which turns out the high voltage at the utility

Figure 7

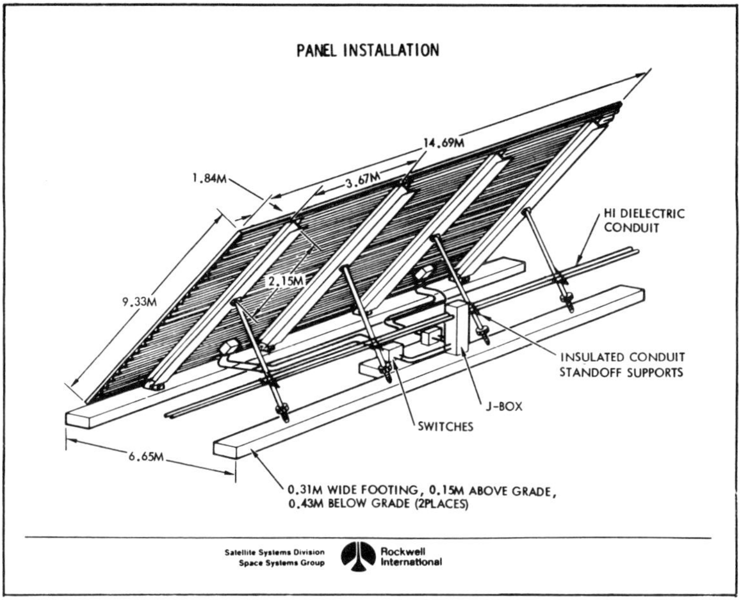

Figure 8

interface. This system covers some 78.54 square kilometers area (48.80 sq. miles). The system would need little maintenance other than an occasional wash down of the panels.

CONSTRUCTING THE SATELLITE ON ORBIT

The important question in determining the space construction system as well as the type of satellite used is the orbit at which it is to be built. It could be built at the orbit where it will be used — at geosynchronous orbit, 22,300 miles up — or you could build it down at low Earth orbit and bring it up to geosynchronous orbit. A number of studies have been done concerning the relationship of the construction site to satellite configuration and it was found that the site made quite a difference.

At low Earth orbit there is a day/night cycle whereas at geosynchronous orbit we are virtually always in the Sun. At low Earth orbit the gravity gradients (variable changes in the degree of attraction between the Earth and the satellite caused by large satellite dimensions) are an order of magnitude greater than at high orbit and that influences the depth of the structure needed to carry the load, since at low Earth orbit it has to be deeper.

The depth and the degree of gravity gradients have an important effect on the amount of attitude control needed to balance the satellite. If we build the satellite at a low Earth orbit and bring it up to high orbit, we will need a lot of propellants to overcome the gravity gradients, and we will need to build the configuration so it will balance around the roll axis. Such complete symmetry creates a problem of inertia balancing.

If the configuration is to be built at geosynchronous orbit it can be much more compact because there is less concern with gravity gradients. We could decide to construct part of it at low orbit, building all of the structure and enough solar cells to power the ion engines to drive it up to geosynchronous orbit, but now there is the problem of going up through the trapped radiation of the Van Allen belt. If we used silicon cells, half of the power would be lost through cell deterioration caused by radiation during the 103 day transfer.

The problem is solved, however, if we do a partial construction at low Earth orbit using the Gallium Aluminum Arsenide cells which would be self-annealing during the trip through the Van Allen belt, cancelling out any transfer damage. Once at geosynchronous orbit the rest of the Gallium Aluminum Arsenide solar cells are installed, the construction of the microwave antenna is completed and operations can be started with the advantage of continuous sunlight and minimized gravity gradients. However, after conducting a number of construction location studies, it was decided that it would be best to do all of the construction at geosynchronous orbit using electric orbit transfer vehicles to carry the construction hardware to geosynchronous orbit.

BUILDING ON ORBIT — BEAM MACHINES

To build a satellite on orbit we need a space facility for the people (about 1000) who are doing the construction to live in, and a construction jig containing beam machines and other construction equipment. Beam machines are nothing new. They have already been built and are used on Earth. The machine builds a continuous set of beams which are joined to make the structure. As the beams are being built the reflectors are put on and the solar cells are installed — all at the same time. The machine builds from one end of the satellite to the other, and once that point is reached, the microwave antenna, which is being constructed separately, is attached. All of this should take about 180 days, working 24 hours a day with the machine putting out beams at less than 1/20th the walking speed of a man.

The machine itself uses cassettes loaded with the needed materials. Material is fed out of the cassettes automatically going through the processing hardware where it emerges as a finished beam. In building a satellite we only have to change the cassettes once during the whole construction period. A de-

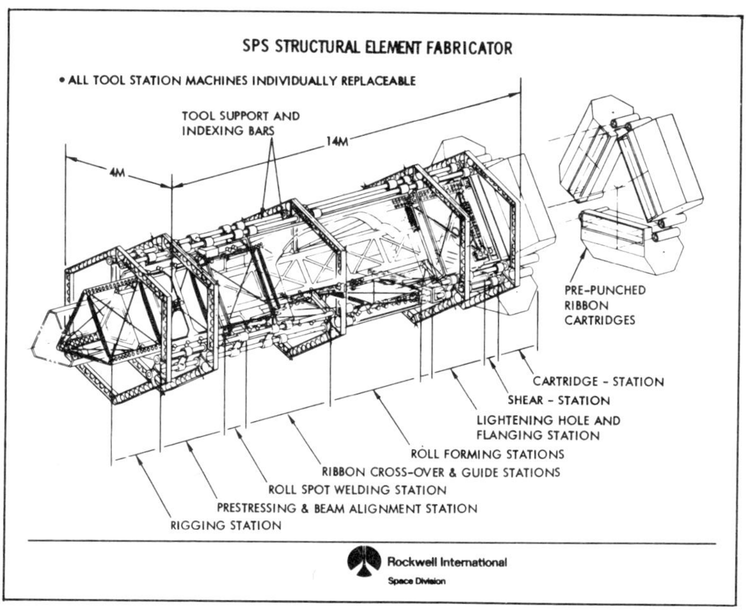

Figure 9

vice called a Manned Manipulator Module is used to remove the empty cassettes and replace them with full ones.

Reflectors are brought up in accordian packages which, when folded and rolled, measure 25 meters in length (82 feet). When open each reflector is 600 x 800 meters (about 1/3 mile x 1/2 mile). Both the reflectors and the solar cell blankets are attached to the structure by means of catenaries (cables). Three Manned Stations are used for this process. At the first station the front of the catenary is attached. Then the beam machine creates the beams and moves along continuously installing the reflectors and the solar cell blankets. Sub-array rolls are delivered to a loader where the Manned Manipulator picks up the rolls and places them into a dispenser which automatically threads and installs them. At the middle deck the long catenary is fabricated and threaded along the side while the attachment process is monitored. On the lower deck there are three Manipulator Modules at each station. They are concerned with the final tensioning of the blanket and the installation of the electrical equipment including the electrical lines, the clamps, the electrical connections and the feeder lines for the Data Management System.

After the antenna structure has been built, the complex job of installing the microwave antenna elements has to be done. The klystron tubes and the wave guides are delivered to an assembly station in the proper amounts where they are automatically put together as sub-arrays and finally assembled into mechanical modules. A crane then puts the completed mechanical module in the webbing of the microwave antenna.

Mass Flow Problems

We anticipate that normally two satellites will be built a year and that will require 125 million kilograms (275 million pounds) of materials to be brought up on orbit each year. Because of these large masses, the Shuttle will have to be replaced by a larger and more efficient vehicle that will be completely reusable and that will bring the cost of launching materials down to $20 to $30 per pound.

One of the vehicles that has been seriously considered for this job is called "Starraker". This vehicle will have a 200,000 pound payload capability compared to the Shuttle's 65,000 pounds. It will take off horizontally, much like a regular supersonic airplane and will use ten air breathing engines to accelerate to Mach 6, turn on its three rocket engines and go into orbit — a single-stage-to-orbit vehicle. Since it is like an airplane and completely reusable, launch costs are cut considerably and turn around time problems are reduced. This vehicle can take off and land at facilities used by current jets.

A transfer crew of about 30 will be stationed at low Earth orbit to supervise the logistics of getting the large payloads from the launch vehicle to the Electrical Orbital Transfer Vehicle (EOTV). The construction crew is transported separately from low Earth orbit to geosynchronous orbit in a chemical rocket-propelled Crew Orbit Transfer Vehicle (COTV). Trip

Figure 10

time for the crew is less than a day; whereas, it takes more than 100 days for the EOTV to transfer materials.

THE PROGRAM

By the year 2000 we expect our first operational satellite to be completed and sending down power. It will take from 1988 to 1997 to construct a 0.5 gigawatt demonstration vehicle and design a heavy launch vehicle that can handle about 200,000 pounds of payload (though it need not be single-stage-to-orbit). Exploratory research is being done now by the Department of Energy to determine the long term effects of microwave radiation as well as interference with the ionosphere, with 1990 set as the date to make the "go" decision. Components will be developed during that time for an orbital testing and space construction site using the Shuttle and Beam Machines. To test the microwave system, however, a linear array microwave antenna will be built on the ground which should be about one kilometer (.6 mile). This will send a beam *up* to geosynchronous orbit (rather than the other way around) to test the effects of the heat on the atmosphere as well as the system characteristics. The beam will be the same regardless of the direction from which the beam is coming. This way the beam can be manipulated and changed with ease on the ground without the complexity of doing it on orbit. By having a relatively small pilot beam satellite and beam mapper satellites at geosynchronous orbit, we can keep costs down and still determine beam control methods.

COSTS

The total development cost of the $35 billion includes all the work that needs to be done to build the 0.5 gigawatt pilot plant. An additional $50 billion is needed to build the first 5 gigawatt unit and includes the complete space construction complex that will be used to build the rest of the satellites as well as a fleet of transportation systems to be used for subsequent satellites. From there on, the cost of an average unit will be $12 billion or $2,400 per kilowatt of power installed. This would represent a cost of about 6¢ per kilowatt hour to the utilities to pay off the construction costs over a 30-year period.

Based on the results of our studies, then, the solar power satellite concept appears to be technically feasible as well as economically competitive, providing an important energy source in some 22 years.

MAJOR STANLEY G. ROSEN, USAF

Assistant for Policy Secretary,
Air Force Office of Special Projects

Major Rosen is a graduate of the U.S. Air Force Academy, and holds advanced degrees in Aeronautics and Astronautics from the Massachusetts Institute of Technology and in Systems Management from the University of Southern California.

He directed development of advanced space propulsion systems for the Air Force Rocket Propulsion Laboratory, and worked on advanced space systems with the German Aerospace Research Establishment. He has also served on the staff of the Science and Astronautics Committee of the U.S. House of Representatives, and as a Project Engineer of the USAF Rocket Propulsion Laboratory. Prior to his present position, he was Spacecraft Manager of the Defense Satellite Communication Office at the U.S. Air Force Space And Missile Systems Organization.

6

THE ROLE OF THE MILITARY IN SPACE

CAPTAIN STAN ROSEN, USAF

Although I am a Captain in the Air Force, what I am going to say today is not meant to represent an official Air Force position. Rather, I want to discuss some of my own personal perspectives on the future of space activities.

The American flag placed on the Moon by the Apollo astronauts is a symbol representing a very unique perspective within the context of the space program. When President Kennedy proclaimed that the United States was going to put a man on the Moon, and Presidents Johnson and Nixon continued this policy, they were representing one of this planet's greatest societies. And when this proclamation was fulfilled — when the Apollo 11 astronauts placed the American flag on the Moon's surface — it symbolized our nation's role in the exploration of space. For space exploration has, until now, been primarily a *national* activity. Moreover, the role of the military in this endeavor has been profound in that our civilian space activities were based on military technology, as well as a military incentive — the space race.

For example, the Atlas missile, developed by the military, has been used extensively in space research and exploration. The very first satellite launched by the U.S. was launched on a military rocket. Many of our civilian space programs have been based on military technology. With the launch of Sputnik I, military necessity brought the United States into the space race with the Russians in an attempt to bridge the gap between the two countries in the field of missiles, technology and national security.

The role of the military in space has therefore been very important in the past and will continue to be so, especially from the point of view of national security. The role of the military is evident in space programs which exist today, and it will have a major role in space programs of the future.

I want to discuss with you a few of our military space programs, to give you an idea of their role in the defense of the United States now and in the future.

TODAY'S ROLE OF THE MILITARY IN SPACE

A major role of the military in space activities today is played by satellites, used for purposes of communication, navigation, and observation. Communication satellites are used for a variety of reasons by the military. IDCSP, the Initial Defense Communications Satellite Program, demonstrated several uses of communication satellites for military purposes and world-wide communications. These satellites were launched on a Titan IIIC, a military launch vehicle which opened a whole new frontier in the space arena. TACSAT, a tactical communications satellite (no longer in operation) was used for precise communication with troops in the field. The United Kingdom and NATO satellites (see Figure 1) are examples of satellites which were built for use by our military allies, allowing NATO countries to cooperate and operate as a military unit.

FLTSATCOM (see Figure 2) and DSCS (see Figure 3), the

Figure 1

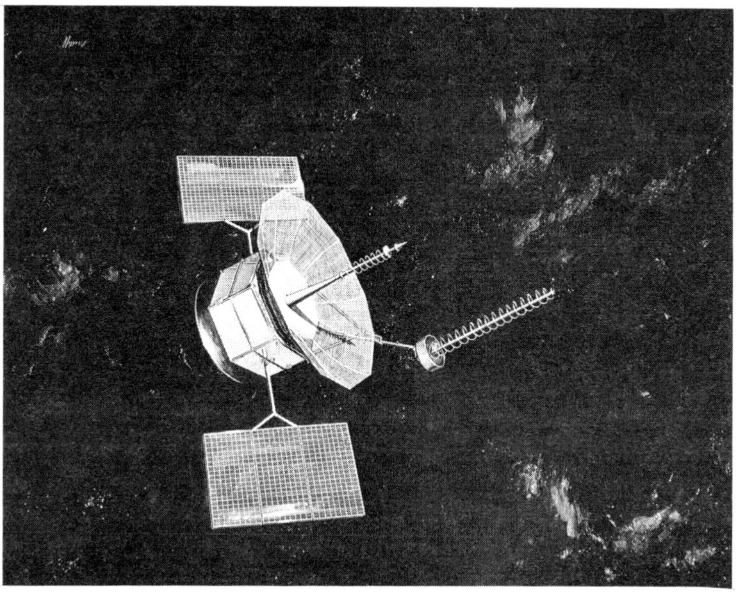

Figure 2

Figure 3

latest generation of satellites, are two other military satellite communication systems. FLTSATCOM is used for communication between the Navy and a number of its ships at sea, providing continuous communication and control. The DSCS (Defense Satellite Communications System) is used for worldwide communication addressing international problems. For example, the White House uses the DSCS to communicate all around the world with our military forces and embassies. The State Department uses it to transmit large quantities of data from one point on Earth to another.

The Phase III DSCS communication satellite (see Figure 4) presently on the drawing board (1978) is probably the most complex communication satellite ever developed. The middle section is about the size of a Volkswagen bus. It is designed to fly on either the Titan or the Space Shuttle, last for ten years in space, and provide the capability for secure, jam-proof communications all over the world instantaneously. Its novel feature, the multi-beam antenna, allows the user to configure a

Figure 4

footprint under the satellite anywhere on Earth, specifying the exact location for desired communications. The boundaries can be as large or small as desired, with virtually any arbitrary shape. This is far better than pointing a dish at someplace on Earth where you want to talk or listen. Communications from other locations may be screened out, so the signals cannot be jammed. Moreover, if communication is desired with friendly forces surrounding a jammer, a tiny circle, or "null' around the jammer would be created, screening out the jammer signal and allowing private communication with others.

The Defense Meteorological Satellite Program (see Figure 5) is capable of assimilating tremendous quantities of data and transmitting these data back to Earth. It provides data used for determining where airplanes should rendezvous for refueling, where the best corridors are to take ships when the weather is clear or cloudy, what world-wide temperatures are, etc.

A satellite system which has tremendous promise for world-

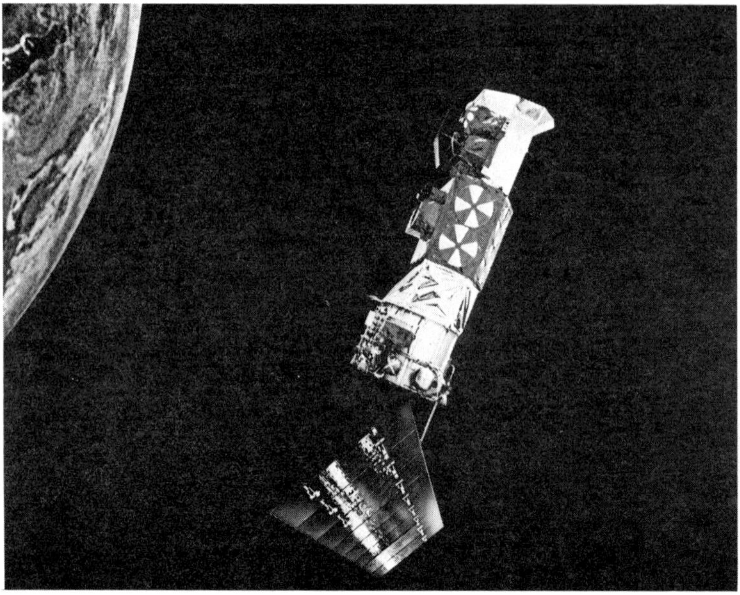

Figure 5

wide military and civilian use is the Global Positioning System (see Figure 6). As a replacement of many present air and sea navigation systems, it will provide precise information (such as altitude, velocity and geographical coordinates) to moving vehicles including airplanes and ships, and to men on the ground. Many users can therefore simultaneously determine precise rendezvous locations and times.

One of the most stabilizing influences in world politics in the last decade has been surveillance satellites which detect missile launches as part of our defense warning system. This capability for early recognition of a major aggression has reduced the possibility of an attack on the United States. Information provided by satellites of world-wide missile launch activity has reduced the tendency to fight, and to use instead increased verbal negotiations. In addition, VELA satellites (see Figure 7) were launched over a decade ago to detect nuclear explosions in the atmosphere. They were launched after the Nuclear Test Ban Treaty was signed, to insure that the treaty was

Figure 6

Figure 7

not violated and thus reduce temptation to violate the treaty. This discussion of military space systems should clarify the point that the major role of the military in space is the assurance of peace on Earth.

In addition, many civilian satellites play instrumental roles in national security. For example, observation satellites in orbit now, such as NIMBUS, TIROS, ATS, and SMS, provide weather information. In some cases, these civilian satellites provide information not available from military DMSP satellites. LANDSAT, an Earth resource satellite system is also vital to national security. These satellites can obtain information regarding the kinds of crops grown around the world, providing insight into world food production. LANDSAT also provides valuable resource location data, such as data which determine the most productive areas for a fishing fleet. The LANDSAT Satellite System is therefore invaluable in terms of assessing factors relevant to the national economy. Finally, many civilian communication satellites carry messages important to national security.

These, then, are some of the ways that the space program is contributing to national security today.

TOMORROW'S ROLE OF THE MILITARY IN SPACE

The roles of the military in future space activity will be to a large extent determined by the influence of our nation's civilian space activity as well as by future scenarios. These factors and possible future scenarios will be discussed first, setting the stage for specific future roles of the military.

FACTORS OF INFLUENCE IN FUTURE SPACE ACTIVITY

With respect to the future of the space program, there are four major factors that I think will influence what we actually do in space, aside from the basic excitement that is generated by space colonies, space industrialization, solar power satellites, etc. These four factors are:

1) Politics.

   Although the political element that emphasizes the importance of demonstrating to the rest of the world that we have the technology to accomplish a given feat, such as going to the Moon, was key to our early adventures, in the future, it will probably be one of the least important factors. In fact, although political reasons have been a major impetus to move into space in the past, they will be relatively insignificant because other factors are so overriding.

2) Science.

   We will continue to explore space simply to gather information about our universe, our origins and other scientific quests. Like a mountain that man wants to climb, to know, to conquer, space is the highest mountain we've yet to climb.

3) Economics.

   The space program is just now beginning to pay off with communication satellites and Earth resources satellites. Unlike the past, economics will be a major incentive to move into space in the future.

4) National Security.

   National security may be the most important reason for developments in space. Although the military has played an important role in the space program in the past, its role will increase to provide the conditions necessary for national and international security, and consequently global stability.

POSSIBLE FUTURE SCENARIOS

We might think of the future of the space program beginning with the Space Shuttle (see Figure 8). One of the first scheduled payloads is Spacelab, an international space laboratory designed and created by the European Space Agency.

Figure 8

With a reusable transportation system, it is probable that within the next 10 to 15 years, there could be significant manned and unmanned utilization of space. There could be a great deal of use of the Inertial Upper State (see Figure 9) plus advanced space tugs to take unmanned satellites and possibly manned platforms to different places in orbit.

There is also great promise for the construction of large satellites in orbit. The technology of integrated circuits and on--board processing will allow us to collect large quantities of data, process it in orbit, and send it to the ground for utilization. There may also be large platforms which consolidate all of the housekeeping functions of power, command and telemetry, attitude control, station keeping, etc., and house a variety of payloads, including those of observation and communication. This would avoid a multiplicity of payloads scattered all over orbit.

It is also possible that such platforms will be shared by a large number of international actors and are likely to stimulate

Figure 9

increased international cooperation. In fact, in the coming years of commercial utilization of space, economic considerations will necessitate international cooperation. The reason for this is that the facilities necessary to support vigorous space activities are going to be relatively expensive so that they are going to have to be shared. Data processing and distribution segments will also be expensive. At present, those space-faring countries which have launching capability are few in number — the United States, Soviet Union, Europe, Japan and China. Other countries who wish to make use of space will do so in cooperation with those countries capable of launch. Third World, underdeveloped nations may also want to use and have access to space resources and scientific discoveries, but without launch capabilities they will be highly dependent on the space-faring nations. Therefore, use of space facilities will require cooperation because of the obvious technical utility and high expense, and international cooperation will be necessary to provide an industrial base for all countries to utilize these space platfoms.

INTELSAT is an example of a commercial cooperative enterprise. INTELSAT has developed a concept of creating world-wide communication via satellites strategically placed in space. Another probable use may be the collection, analysis, and dissemination of data of Earth's resources, such as weather information.

FUTURE ROLES OF THE MILITARY IN SPACE.

The future roles of the military in space include the following:

1. Force Multiplier
2. Support of National Policy
3. Protection of Assets in Space
4. Counter Force

1. Force Multiplier.

The utilization of space as a Force Multiplier allows for a small number of forces to do many more things than they could if they didn't have the space segment. For example, direct communication between the President and the American embassies, the general and his troops, and continued observation of Earth activities are very beneficial to the military. They allow for more control of what is happening to any group in any environment and in any particular combat zone. Satellites will be used to provide precise weather and navigation information, providing the best possible use of available forces. Observation platforms may track activities to prevent intrusion into areas attractive to a military enemy. Observation and the subsequent control of specific areas on the ground will therefore be a tremendous military utility, acting as a Force Multiplier.

2. Support of National Policy.

There will be continued observation of Earth activities for

treaty verification purposes, leading to greater strategic stability. Global activity observation will warn against nuclear events and other attacks. Thus the military will provide the kinds of information necessary to support national policy and in some cases, to create it.

3. Protection of Assets in Space.

A frequent question is how vulnerable our satellites are to an enemy attack. Most of our satellites were not designed to resist any kind of interference and most have not been tested for their vulnerability. Laser weapons in particular are a potential threat, in that lasers are able to project energy in extremely narrow beams, creating highly destructive energy that can be directed at very small, precise target areas from a distance. Therefore, the military will have to develop the technology to protect space assets, including civilian satellites, just as it has been called upon to protect Earth assets around the world today. In the coming years we will have significant investments in space systems; these investments must be protected.

4. Counter Force.

The military's potential counter force role is one which is frequently misunderstood. Counter force means a force or weapon against a certain military target, as compared to counter value which means force against a civilian target. Of course, weapons of mass destruction are prohibited from use in space. Nevertheless the time may come when limited space weapons emerge.

For example, high energy lasers in space could drastically change the balance of power. When lasers are placed in space so that every location on this planet is placed continuously in a target area, rapid delivery of nuclear explosives may be prevented. The nation possessing such a system would provide interception of enemy ballistic missiles during their boost phase with relatively small amounts of energy and relatively little destruction. The precision capabilities are of greater importance than the release of energy. This system is therefore a weapon system which in itself is *incapable* of mass destruction. Balance of power is therefore maintained without the need for mutual destruction.

OUR GLOBAL FUTURE

Like the commercial consortia endeavoring to promote commercial utilization of space, similar consortia based on military interests, such as NATO, are likely. The NATO satellite system is a model of what the future holds because we are going to form blocks of countries coming together for their own security, cooperating together with the use of military space systems. Thus, both commercially and militarily, space utilization may well lead to more international cooperation and perhaps increased global stability.

NICHELLE NICHOLS

President, Women In Motion Inc.

In her initial film experiences Nichelle enjoyed many roles, but to millions of fans all over the world Nichelle Nichols is "Lieutenant Uhura" the Communications Officer aboard the Star Ship Enterprise in the television series "STAR TREK", and now the motion picture, "STAR TREK-THE MOTION PICTURE."

As a member of the board of Directors of the National Space Institute, she delivered a speech "New Opportunities for the Humanization of Space" which called attention to the lack of women and minorities in the Space program. NASA officials subsequently invited Nichelle to join them in their astronaut recruitment program for the new Space Shuttle. Nichelle contracted with NASA through her own consultant firm, Women in Motion Inc., and played a key role in bringing women and minorities into the Astronaut Corps.

In recognition of her contribution to inspiring, and informing young people of their Aerospace Heritage, the American Institute for Aerospace Education honored her with their 1979 Woman of the Year: Friend of Space Award. The American Speech-Language-Hearing Association, Washington D.C. has appointed her as there honorary chairperson for their National High School Essay Contest on Communications.

Currently, Nichelle is rehearsing for a new record album and is co-writing a theatrical musical called "Ancestry," which tells the story of a female astronaut on a space colony. She also leads the Space Cadets of America, an organization for young people interested in space and space careers.

7

THE FUTURE IS NOW

NICHELLE NICHOLS

When I was chosen for the wonderful role of Communications Officer in the television series "Star Trek", I had no idea it would have such a far reaching effect on my life today. By a quirk of fate, I literally became involved in the motivation and recruitment of women and minority groups to apply for astronaut status. As a consequence, I have felt like the once and future astronaut.

At the time I portrayed the character Lieutenant Uhura, a rather strange thing began to happen. A sort of fluttering feeling kept reappearing in my left ear and I was thoroughly confused as to what it was. As this feeling kept recurring, I somehow felt that Uhura was watching over my shoulder. When an actor creates a character, you breathe life into it, and so I did when I created Uhura. And, in effect, she had an identity crisis. I was either taking her home with me, being her, or brutally stashing her away and denying her existence until work the next morning. It seems that by the very job I did in creating her, the more she wanted to exist.

A year after the show was cancelled in 1970, a grass roots effort to have "Star Trek" conventions began in this country and spread throughout the world. It seemed that people wanted to keep the dream of Gene Roddenberry's universe alive. They would convene to invest their love and interest in what was not just a television show or wishful thinking, but an actual blueprint for reality. However, it wasn't until the 1975 "Star Trek" convention in my hometown of Chicago, that I began to have the slightest inkling of how I would become involved in the "Star Trek" universe reality.

It was possibly the largest "Star Trek" convention held anywhere at any time. It was also the first time NASA sent a scientist, Dr. Jesco von Puttkamer, to give a serious presentation on the long-term space program. He was scheduled to speak the first night of the convention, and I told my security force I wanted to attend his lecture. They were concerned about my safety but I explained that "Star Trek" fans were very special people and I could handle them.

I arrived in the lobby incognito as Nichelle Nichols, but I was swamped by hundreds of fans who simply wanted an autograph or to say hello. I explained to them that I had just arrived in town and I cherished their friendship, but that night I just wanted to join them as a "Star Trek" fan to attend the presentation by a NASA scientist to find out what was currently going on in the space program. This message quickly filtered throughout the crowd. Just five minutes before, I might have been mobbed, but there I sat in the auditorium with all of those wonderful people, sharing a common interest together.

I was struck at that moment by the beauty, magic and essence of the "Star Trek" universe and what it was bringing forth. I was totally thrilled at what was taking place and felt the wonderful intimacy shared throughout the audience. The lights dimmed and Dr. von Puttkamer began to talk about the long-range plans of NASA. I was flabbergasted, because up until that point, I really knew nothing about space. As far as space was concerned, I had been creating and projecting a fantasy character like Buck Rogers. Never had I conceived of actually going into space, or for that matter, what our future might hold in store.

Suddenly I knew that Uhura's world isn't far away and the little tickle in my ear grew stronger. I became aware that it was Uhura's voice, sounding very much like my own, as though she exemplified my alter ego. I heard her whisper, "I finally broke through — the future is now! I want my world to live. You created me and you're responsible. If you don't do something now, my future, our future, and the "Star Trek" universe may never exist. The dream may never become a reality!"

It then occurred to me that the highly capable, rather stoic Uhura was worried. Was this my conscience? I had tried to infuse in her character all of the qualities and strengths a person of her station would require to be chosen for the post of Chief Communications Officer. Uhura was responsible for all contact with other life forms and was an extremely important member of the "Star Trek" mission. It occurred to me that because I created her being, she must also possess my genes, qualities and soul. I had created Uhura in the now, but in her universe still 200 years hence, she could very possibly be my great-great-great-great-granddaughter. Realizing this, I felt a responsibility not only to the "Star Trek" philosophy, but also to the lives of our future generations. As ancestors of future generations it's what we do on Earth *now* that will judge us worthy. In future times our posterity will look back to us and say, "Thanks great-great-great-great-grandparent. Job well done, my wonderful ancestor. What a beautiful world you've left me!" And so I discovered that truly, in the world of Lt. Uhura, "the future is now!" From that decision, I have become very involved and tried to learn as much as I could as a layperson about the real space program.

This role for Nichelle Nichols became possible when NASA began to search for a person who could reach people and convince the public that NASA no longer desired an all-white male astronaut corps. They asked me who I thought was best qualified. Advising them to find someone who was visible, credible and convincing, I gave them a list of various public figures and celebrities. NASA then said they wanted Nichelle Nichols. I told them I was a fictitious character, but they re-

Nichelle with Isaac Gillam, Director of NASA Dryden Flight Research Center

plied that I had been in the hearts and minds of many, many people. It seemed that I really had no choice but to accept the job.

My life has been incredibly enriched because of my participation in the subsequent four-month tour. I travelled across the entire country lecturing to various ethnically oriented and women's professional organizations, and scientifically oriented universities and conventions. I also appeared on television talk shows and commercials to disseminate information about the space program. The wonderful, infinite diversity and combinations conveyed in "Star Trek" became a reality for me. I recruited not only women and minorities, but also white males to apply for astronaut status.

As I spoke to people about the space program, I discovered that very few know anything about it at all. I also found that many are alienated by the unknown realms of the universe and others are frightened by its vastness. We should have positive expectations about what is in the universe, not fears and dreads. We are made larger with the realization that we're not Earthbound, and that our acceptance of the universe offers us room to explore and extend outward. It's like being in a dark room and imagining all sorts of terrors. But when we turn on the light — technology — suddenly it's just a room where we can stretch out and explore. If the resources here on Earth are limited, they are not limited in the universe. We are not constrained by the limitations of our planet. We are only limited by our minds. We must therefore begin to realize that growth is enriching and ennobling.

To this end we are fortunate that humankind are dreamers. It's the dream in all of us which makes a feat possible. The initial vision enables us to discover the means by which to get there. If we tone down the dreamer we also tone down the discipline. A nice push-pull theory evolves from this idea. Our desire to transform the dream into a reality is the pull. The push is our decision to create the means to obtain that goal. We don't have the push first; we begin with the dream as the pull. We don't develop the technology and then say "Well, now that I've developed a terrific propulsion system, what am I going to do with it?"

We've been gazing at the Moon since time immemorial, trying to figure out a way to get there. This dream was a pull for thousands of years and someone finally said, "a different kind of propulsion and I'm going!" That someone was John F. Kennedy who declared we would travel to the Moon in a decade. The edict was out. Humankind said we have a dream, a directive, and we're going. And we went!

So our scientists are now very busy in their marvelous ivory towers developing the technology which benefits and raises the quality of life here on Earth. But nobody really knows anything at all about their work. All the layperson really knows is that some guys walked out on the Moon, and how do you follow an act like that? The nonprogressive claims it's only greatness rearing its head again. But there are many who have truly wanted to learn about it, yet felt disenfranchised, because they didn't see the benefits the space program could give to them personally. And honestly, we all want to know, "What's in it for me.?"

The fact is, space belongs to all the people, and we have a responsibility to learn and educate others about it. We have a wondrousness here — we can walk in the universe — but no one is teaching our children that this greatness is a natural extension of their natural genius. The development of technology is a natural extension of our God-given talents, and what we think we can do, we will do and better.

So the time has arrived for us to deliver our own edict and say that we will move out into the universe. Everyone will be involved. For every discipline here on Earth, we will correspondingly take it upon ourselves to incorporate a wholistic environment in space. We'll have to pool all of our resources and talents to expand our human potential. We need more scientists and engineers than are currently available. Women and minorities, a vast untapped resource, must be encouraged to enter non-traditional roles, venture forth and reach out to one another to make the dream possible. We need the special talents and abilities of persons who can contribute not only in the scientific fields, but in the arts as well. The creative minds of artists, writers and musicians have always expressed our universal options and dreams.

Yet we may still be faced with the question, "Why space?" People want to know why we should spend our energies, resources and even our emotions to further the space program. NASA spinoffs are benefits which everyone can observe, including pacemakers, microwave ovens, digital watches and improved computer technology. Significant breakthroughs in medicine have been developed because of the space program. Observation satellites have contributed great amounts of information about Earth resources and the field of oceanography. Communication satellites have created a global network of communication systems. The answer, therefore, is that there is no limit to the human mind's creativity and capability to develop the technology we can utilize for our growth.

The marvelous advent of the Space Shuttle is bringing us into a whole new era of transportation, scientific exploration and experimentation. We may discover replacements for our rapid depletion of natural resources and may need new environments for habitation if world-wide overpopulation continues. Certain types of manufacturing can be done far more inexpensively in space because of the vacuum and zero gravity conditions. Going into space can therefore be a good financial investment generating solid returns.

Unfortunately, after we've accomplished any exceptional feat, there are some who always ask, "What will we do for an encore?" For example, after building a machine called "Viking" that can travel 250 million miles, land on another planet, Mars, use its equipment to investigate materials and transmit the information back to Earth, many members of the press were not too impressed, claiming that we didn't find life. It's not really important what we did or didn't discover. The true importance lies in the accomplishment we made and the contribution it made to our current and further endeavors.

After writing a letter to express my disenchantment with the negative news coverage on a television station, one newscaster said that perhaps we do knock down our greatness too often, and should concentrate on our positive accomplishments. I had written the letter as an interested citizen to tell them that it was *my* universe they were speaking of, and that although they didn't feel it was worth exploring, I most certainly did. The media reaches millions of people including many, young, impressionable minds. I told them they just might be stunting youth's ability to imagine, dream creatively and perhaps someday project those dreams into future realities, now only imagined.

Some critics say we should spend our money solving problems here on Earth, such as cleaning up the ghettos. But not going into space doesn't mean we are cleaning up the ghettos. This is a false argument. Realistically, we're not throwing our dollars away for space exploration, we're spending them here on Earth and are bettering our lives here because of it. We are also making breakthroughs in studies of lengthening life expectancies by finding ways to help human beings withstand alien environments.

Another common criticism of space development is the area

concerning killer satellites. They may exist, but we don't necessarily have to use them, and they certainly don't justify putting an end to the whole space program. We will always have a defense program, but the important thing is to keep a balanced perspective and concentrate on our technology as a positive tool. A politician once told me she was against the Space Shuttle program because there was a possibility NASA would explore defense systems with it. I pointed out the numerous, positive explorations the Shuttle would also conduct. If we want to limit the area of defense as responsible voters, fine. But why limit the whole Space Shuttle program in its entirety? It's like throwing out the baby with the water. The more we learn and grow from this perspective, the less we will need to defend ourselves by dealing with each other in childish ways.

The intangible benefits of going into space are less obvious than the practical reasons, but are just as important. Our trek into space is the expansion of the human spirit. Humankind has always reached for its highest ability; its greatest potential. From our inception, we crawled up out of the oceans, climbed up the trees and now we're extending ourselves toward the stars. As children have to leave the security of family and home life to insure growth into mature adults, so also must humankind leave the security and familiarity of Earth, to reach maturity and obtain the highest attainment possible for the human race. The growth which will result from our journey into the realms of the universe is growth in self-awareness and of our place in the universe. Our minds are freed from their restraints as we experience the infinite. Knowing and feeling inside that the universe and ourselves are infinite, we will grow and learn to be worthy of greatness.

To really reach out and grasp the dream humankind has sought, is both possible and realistic. If we think the human race has this potential, then we must gather up our forces to call the future into being. It is with our personal involvement, however, that we can begin to transform our dreams into future reality. We can begin by informing universities and colleges of these possibilities, motivating them to offer courses which relate to the subject matter. Space industrialization courses are only now being offered at educational institutions. Primary and secondary school teachers who learn about the space program can apply it to their course curriculum and children's eyes will sparkle with interest. Chemistry, engineering and computer technology will enable students to obtain the foundation to work in the aerospace industry.

We need to investigate the various career opportunities which exist, and the specific disciplines that will be required for the future. Law will be an extremely important area. What will our form of government be in space? Where are our allegiances? Will we retain our national citizenship status or will we be colonists when we move to space habitats? What will be the identity of the first child born in space? These are only a few of the questions which will need to be answered in determining the rules and regulations of our lives in space.

We need more responsible political leadership and we can begin by writing letters to the White House and government legislators. We should let our congresspersons know of our interest in the space program and find out what steps are being taken to insure a viable, growing space program. It's the responsibility of intelligent citizens to avoid becoming the victims of the status quo by making our needs known, by organizing to present our desires as a group.

We should contact the aerospace industries and find out how we can become involved in the many community outreach programs they offer. They desperately need people who are knowledgeable about many areas of space to educate and inform the public. The National Space Institute is a marvelous source of information, consisting of people from industry, government and the private sector who are interested and involved in the space program. Members are kept informed about their diverse activities, including efforts to form legislation to further space efforts and keep the dream alive. Forming and joining space-oriented clubs also provides a vital forum for information exchange.

The involvement of everyone in every branch of learning is essential to keep the dream of our future alive. That great man Gene Roddenberry put forth a vision where all beings are created equal in a wondrous world of infinite diversity and combinations. The inhabitants of his created universe respect and accept one another's differences, genius, and technology as an extension of themselves. To his creation I would also add that we should revel and delight in those infinite diversities and combinations, accept the challenge and become involved. For as I, Nichelle Nichols, and as I, Lieutenant Uhura, do truly believe, the future is now!

KRAFFT A. EHRICKE

President, Space Global, Inc.

Dr. Krafft Ehricke started his space and rocketry career in 1942 as a research engineer on the German V-2 rocket program. He came to the United States in 1945 and until 1952 worked on the U.S. Army missile and rocketry development program.

From 1952 to 1954 he was Assistant Project Engineer on the long-range glide rocket studies for Bell Aircraft. Subsequently he joined Convair to work on the development of the Atlas ICBM. While at Convair he conceived, designed, and directed the development of the world's first oxygen/hydrogen upperstage rocket, the Centaur, which led to the choice of cryogenic upper stages for the Saturn Moon rockets. From 1959 to 1962 he was Chairman of the NASA Advisory Committee on Electric Energy Systems and Solar System Exploration and subsequently directed many studies for NASA involving advanced launch vehicles and solar system exploration. From 1965 to 1977 he worked for Rockwell International as Assistant Director of the Astronics Division and later as Chief Scientific Advisor for Advanced Programs, Space Division, at the Los Angeles corporate level.

He has published hundreds of articles and many books. His two volume series, SPACE FLIGHT, is used in courses on Celestial Mechanics throughout the world. His forthcoming book, THE EXTRATERRESTRIAL IMPERATIVE — FROM CLOSED TO OPEN WORLD, deals with the logic of space exploration and industrialization.

Among his many awards, he was presented the 1st Guenther Loeser Medal in 1956, he was named to the International Aerospace Hall of Fame in 1966, and in 1972 he was given the Kurt Laskowitz Aerospace Engineering Award. He is listed in WHO'S WHO IN AMERICA AND AMERICAN MEN AND WOMEN OF SCIENCE.

8

THE EXRATERRESTRIAL IMPERATIVE

KRAFFT A. EHRICKE

I remember Vice-President Agnew's talk on the eve of Apollo 11. The talk was about how we were now ready to go to Mars. Here was a nation that had just spent 25 billion dollars on a project. We had just received the first pictures back from Mars, showing a heavily cratered surface. It looked as if Mars was just another moon. The President's Scientific Advisor, Jerome Wiesner had just announced rather perfunctorily that a manned Mars mission would cost about 200 billion dollars. Unfortunately the manuscript delivered by Mr. Agnew showed no social sensitivity. There was no mention of any returns for the American people from Apollo, or a Mars mission. That talk choked public enthusiasm and eventually provoked vigorous criticism of the space program as socially and economically "irrelevant," despite the fact that neither accusation is true. Research and exploration is an investment in our future. I was therefore disappointed that the Apollo missions were not the high point of a socio-cultural experience I naturally expected them to be.

At that time, I said to a colleague, "Let's see whether it is really true that this is only the modern equivalent of Roman Amphitheater games played by modern superpowers." Did we give ourselves just a race into space and to the Moon while people are starving? Is there a humanistic relationship between our budding space capability and the state of mankind? Is expansion into space a natural part of evolution or a cancerous growth in a sick industrial world, as cultural pessimists, environmentalists, and limits-to-growth people have claimed? If space development is a socially and economically meaningful part of our present development, then we have a justification for space which goes beyond any temporal merits of the political race that originally motivated President Kennedy.

I found a definite evolutionary connection from which fundamental justifications and an evolutionary strategy of expansion into space can be derived. I called the evolutionary connections the Extraterrestrial Imperative, for reasons that will become apparent in the future course of this lecture. I wrote papers on the subject, made presentations to NASA and testified before Congressional Committees. But the interest in space waned as fast as it had waxed before. No foundation or in depth understanding had been laid and there was no interest in one. America's space program had begun as a reaction to the Soviet Union, not for its own sake. The concept of space industrialization was not discovered officially until 1975/76. Now we have lost a decade of possible progress in manned space flight and preparations for productive space utilization.

THE EVOLUTION OF LIFE

Technology is life because without the functional use of matter according to natural laws, life would not be possible. Technology has therefore been with us since the beginning. Life, however, has had apparent limits to growth before; evolution is the advancement of life's technological capabilities to overcome those limits.

Earth cannot sustain life without help from the outside for

any length of time. At the inception of life, the primoridal atmosphere was composed of methane, ammonia, and similar substances, but not of free oxygen (it does not really matter whether the original amino acid formed out in space and came floating down, or whether it formed in the oceans of the Earth).

When life—i.e. cellular organisms with reproductive capabilities (genes)—arose on this planet, it existed on the basis of chemosynthesis (chemical synthesis) through a metabolic process that more closely resembled fermentation. For its energy source, it relied on "fossil" substances, that is, abiotic carbohydrates (in other words, amino acids and other carbon-hydrogen molecules that had been formed by inorganic processes such as lightning, solar ultraviolet radiation, volcanic eruptions). The abiotic carbohydrates existed in the original oceans and their processing by single-celled organisms called autotrophs required only a very primitive metabolic technology. Thus, the DNA of those organisms probably did not contain more than 2 or 3 bits of information. Exhaustion of the limited supply of abiotic carbohydrates brought on the First Great Crisis of life. Those early single cell organisms ran out of basic nutrients and became the "soup that ate itself up."

The question then was, where can more energy be obtained? Matter was there, but Earth, being a small, highly oxidized planet, was in a state of high entropy. In other words, there was not much free energy around. The result was a desperate fight for survival. Apparantly many of those autotrophs did unbelievable things to try to stay alive. They began to process more difficult inorganic matter such as not completely oxidized silicone or metallic compounds. Others learned to squeeze energy out of the waste of autotrophs still feeding on remaining carbohydrates and to consume dead autotrophs floating around. This learning process, triggered by the search for energy, pushed their development and growth. It pushed the development of new enzymes and more complex chemical regulation processes. It improved the hereditary matter of the survivors, because more information was needed to carry out those more complex chemosynthetic processes. The enzymes are catalysts; necessary tools for the metabolic technologies employed.

In the course of this growth, some autotrophs learned to jump other autotrophs. They gave up on the hard job of seeking a living out of the inorganic, primordial environment and became parasites — called heterotrophs. As the "herds" of autotrophs began to shrink, the fight for survival began to break out among the heterotrophs. It caused a further differentiation — heterotrophs living off other heterotrophs.

So, this continuing and rising pressure increased the information storage capacity and improved the metabolic technology of the organisms. But parasitism only amounts to a distribution of scarcity. It does not add to the wealth producing capacity; and the remaining autotrophs — the only ones capable of introducing energy wealth into the biotic system from the "outside" (i.e. primoridal matter) — made less and less progress. All their technological improvements were along the same basic lines: chemosynthesis. It became a matter of diminishing returns and, therefore, was to no avail in the long run.

Living organisms are a work-performing system. They must have access to an open energy environment and an open material environment, or in other words, an *open world.* Matter can be recycled, energy cannot. Primoridal life had to get access to a new energy source.

Earth does not have a suitable planetogenic source for organic life to handle. To the limited capabilities of biological technologies, there is only one other energy source suitable — an extraterrestrial one, the Sun. At that state of life's evolution, the Extraterrestrial Imperative asserted itself for the first time as a driver for survival of life on Earth. To use solar energy, life did not have to leave Earth. The energy was flowing to Earth in the form of solar radiation. But life had to develop the complex technology needed to use solar energy.

It was the dawn of the first industrial revolution of life on this planet. Among the many types of autotrophs, one type evolved an enzyme that could operate photochemically — that is, it was able to assimilate solar energy into its body processes, as opposed to just reacting to solar energy. Perhaps other autotrophs could react to light, such as getting out of the sunlight or seeking illuminated regions. But they were not using it the way this photo-autotroph was. The photo-autotroph applied photochemical effects as a form of nourishment by converting solar radiation energy to chemical energy that could be stored in the organism. It was the ancestor of the chlorophyll molecule.

The chlorophyll molecule is an extremely complex technological development. It industrializes life. The chlorophyll molecule carries out a *mass production process* of the basic staples of life — sugars and starches. Of course, the result of this engineering advance was a waste product that forced life to take the next step. As bio-industrialization progressed, it "polluted" the "untouched" environment of the original Earth. The waste product was free oxygen that spread around the globe.

Free oxygen was a poison for the life that existed in that environment. The oxygen was released gradually, and so there was time for life to adjust to the new environment. But many of the original species could not adapt and perished as the free oxygen spread and permeated the water and the atmosphere. This eventually became an advantage, as it caused the ozone layer to form, which protects exposed life from harmful radiation, enabling it to eventually emerge from its protective water world. But life's technology was not yet ready for such a giant step. To survive, organisms that were not able to photosynthesize had to learn to use this waste product, oxygen, and develop the technology of oxygen metabolism. Life was on the way of becoming globalized as well as industrialized.

Photosynthesis solved the energy problem. It did not solve the material problem as carbon dioxide was not plentiful enough to last forever. Prior to oxygen metabolism, life began "eating up" the carbon dioxide in the process of creating free oxygen and assimilating carbon and hydrogen. Life had to evolve a technology that could handle the oxygen and convert it back to carbon dioxide, thus closing the material cycle and assuring a carbon dioxide supply for the indefinite future.

Of course, there were various stages of oxygen metabolism. Only plants can perform both photosynthesis and oxygen metabolism which makes them the "autotroph descendents" of today. Animals can only perform oxygen metabolism. Their food chain begins with the parasitic use of the solar energy stored chemically in the plants they eat. This makes today's animals "descendents" of the heterotrophs. Oxygen metabolism, therefore, is a secondary process. Photosynthesis was the big engineering leap forward. It triggered everything else and put life on a permanent basis. However, while essential for survival, photosynthesis created a formidable disturbance, as Man does today. Oxygen metabolism adjusted life to the disturbance, enlarged its environment and restored equillibrium, in a new and higher order. This is growth.

Thus, what some today almost worship as "natural environment," is in reality a bio-industrialized environment created by the destruction ofthe earlier "natural" environment of our planet. Our environment is the result of change, and change is the most natural process in nature. Only *we* do not have to all-out destroy the present bio-environment, since we can expand our metabolic industrial capabilities beyond Earth; something earlier bio-industry was not able to do.

With the evolution of oxygen metabolism, the evolution of progressively higher organisms, and finally Man's evolution, not only DNA information capacity had to be increased enormously, but also growing sensory capacities. Sensory information input rates increased, and better data storage and processing had to evolve. This was the cerebral evolution.

The cerebral evolution actually led to a new type of metabolism which I call Information Metabolism. By this term I mean the ability to take individual bits of information, take them apart, and put them together again into coherent systems of knowledge and capabilities. This is not necessarily the only form of intelligence, because a number of higher animals also display it. But it is a particularly powerful and effective form of intelligence; just as some 3.6 billion years ago, photochemical processing became a particularly powerful and effective form of utilizing light. It forever changed the face of Earth and gave life a gigantic new evolutionary potential.

Information Metabolism probably arose in response to a particularly difficult environment — the crowns of trees in a forest. There you cannot move safely as on the ground, or fly freely as in the open sky. The jump from one tree to another or from one branch to another, as an ape or monkey does, is a vast complex process. "Three-dimensional" observation is needed. A lot of information must be acquired and processed quickly, such as where the target branch is located, how far away it is, how strong it is, how much weight it might support, the effects of shadows and perspective distortions, and so forth. This information then has to be translated into mechanical motion. That is where cerebral functions naturally evolved into a very rudimentary form, or perhaps were a direct forerunner of Information Metabolism. It became a metabolic breakthrough in recent evolution, comparable only to photosynthesis in evolutionary significance.

When new metabolic capabilities are developed, the consequences include new capabilities, new processes, and new waste products. These waste products may not be adapted to the natural or existing environment, thereby creating conflict. With the first step in evolution, the first Earth was radically altered. Now, present life cannot do without oxygen metabolism or without chemosynthesis. With the rise of Information Metabolism, the human life form has gone one step beyond that, and again we have conflict. The conflict is always at the same interfaces: energy, production/processing technology, waste products, raw materials, and environment/processing space. The basic parallel between the First Great Crisis and the elements of the threatening Second Great Crisis are striking and demonstrate the general validity of evolutionary principles.

The biosphere evolved into a multi-environmental system that is at the same time industrial processing space and living space. Everything produced in that system serves as food for something else. Yet the very fact of that production and mutual use is what makes life so pleasurable. Things smell good, taste good, and the atmosphere is pleasant. Animals too, react to it. Therefore, while nature is cruel, it is at the same time pleasant. Processing to eat or be eaten, to slaughter or to run a steel mill or a factory is not the most delightful thing, but it is necessary. Our human industrial processes enrich our lives beyond the wildest dreams of earlier generations, yet they also give us waste products.

You must also realize that the decision to industrialize was made years ago. To go back or to stop in midstream as we develop progressively better technologies would threaten destruction of humanity and the Earth's biosphere. We have to go forward. We have to improve. We have to grow and grow up. Life as we know it is an industrial process. Man continues this natural evolution. To reject the technology of life is to reject life itself.

If you compare the biosphere that holds all life on this planet to a house, then the organisms of ecological niches may be compared to pets in the household. Maximal adaptation is the maximum point of evolution. Those that come up with a new umbilical metabolism, such as photosynthesis and Information Metabolism, are not pets, but rather children of that house. Their ultimate state is not maximum adaptation to the conditions of the house, but the ability to outgrow it and establish new conditions. One day the child will want to leave the parents' house and establish his or her own family, his or her own career, his or her own "world." They would resent it if the parents would enforce conditions that would prevent them from doing that. Leaving the childhood home does not mean rejection of the parents, but rather shows natural growth and evolution. As humans, then, it is our task to grow out of our house, Earth, not adapt to it at the expense of our growth potential.

As the parasitic embryo eventually becomes mother's protector, so can the emancipated Human become preserver and protector of plants and animals, because he has the knowledge and the power. Today, when we have a drought in Africa, helicopters are employed to fly antelopes and other animals to a place where they can get water. The paleolithic hunter could have done nothing to save the animals. Even if he had the idea, he didn't have the means or power. But we have this power because we exercise our Information Metabolism. We are not trying to totally adapt or blend into the biosphere, for we have no ecological niche. Yet, we are nature too! We are another evolutionary form of nature, specifically of animate nature. As such we manipulate the environment, and find ourselves in conflict with the environment. It is part of our growth, part of our natural evolutionary process.

To think that we could stop growing could be compared to an imaginary embryo that is in its sixth or seventh month and has decided to stop growing in order to survive in the womb. The womb is the only environment known; all others are feared out of ignorance. Assume, however, that the embryo is very intelligent. It has kept records since the third or fourth month. It extrapolates from its statistical data to the eighth or ninth month. It sees environmental conditions in the womb getting precarious going into the tenth and eleventh months. It decides this growth is impossible, so that it had better stop growing in the fifth month or the sixth month before a catastrophe occurs. What it doesn't know is that in the ninth month a change will take place. It will be called upon to activate new technological capabilities and exist in a new frame of reference. It will have to use its own lungs, its own stomach, and there will no longer be a placenta provided by the "mommy" who conveniently carries all the wastes away. It will be born into fully human life by emancipation from the womb.

"Mother" Earth and her latest children, humanity, are at that same point now. Our new frame of reference will be the environmental enlargement beyond Earth. Now that we possess the necessary technology, we can "breathe" and live beyond Earth, outside the womb of the biosphere in which we grew up. That is where our direction of growth lies. We *must* again reach out beyond Earth. The Extraterrestrial Imperative asserts itself again as an evolutionary driver. But the growth is achieved not simply by moving out to space, but rather by actually combining Earth and space into one greater system — a system that is a technological/industrial civilization based on terrestrial, space, lunar and other environments.

Such a system will be far more stable than one that is based only on Earth. Exclusively terrestrial systems are not very stable. They were abandoned some 3 billion years ago. So while it is theoretically possible that we could solve all our problems without going anywhere, it is not probable that this approach would be successful. More likely, it would either destroy our evolutionary potential or our biosphere — possibly both. We need to enlarge our environmental base.

Now, if we are to move into that new environment, into space and toward exoindustrial utilization, we need to put that move into prespective. We need a broad, thoughtful approach to the utilization of all our environments, including the socio-economic reasons as well as scientific and technological reasons. We must apply our Information Metabolism.

EXOINDUSTRIALIZATION

The larger environment of Earth and space can represent an integrated unit in which processing space and living space can fit together in great harmony (Figure 1). This development can lead to a stabilization of terrestrial population, a lasting resource foundation for our industrial civilization and a progressively improving correlation between terrestrial and extraterrestrial environments, processing space and waste disposal. Exoindustrialization promotes the completion of industrialization on a global scale — meaning the changeover from pre-industrial to more productive industrial societies (with their appropriate agricultural differences) while the terrestrial environmental burden remains within manageable limits. The environmental base is widened to a terrestrial-extraterrestrial complex. In industrial societies, however, the socio-economic environment favors smaller families, as experience has shown repeatedly. This fact provides the foundations for the above claim that space industrialization will promote the stabilization of the terrestrial population. This is one of the three key items that constitute the long-range social benefit potential of humanity-oriented space industrialization

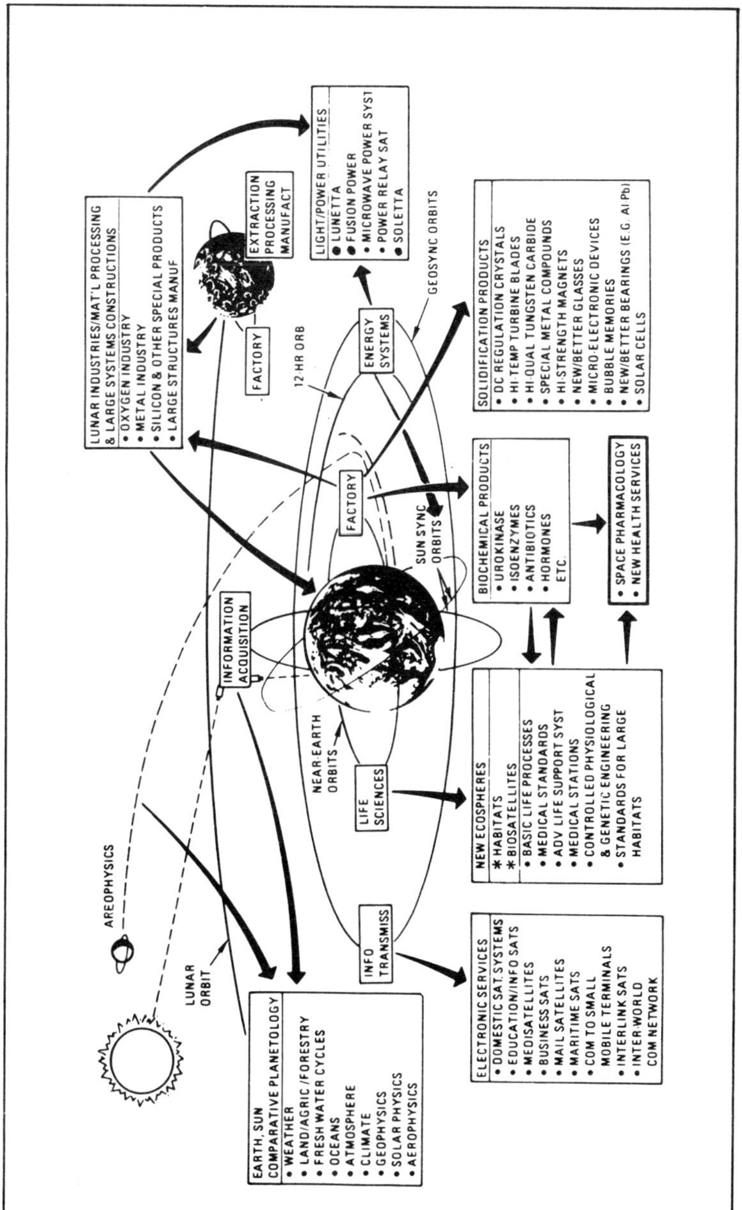
LUNAR INDUSTRIES/MAT'L PROCESSING & LARGE SYSTEMS CONSTRUCTIONS
• OXYGEN INDUSTRY
• METAL INDUSTRY
• SILICON & OTHER SPECIAL PRODUCTS
• LARGE STRUCTURES MANUF
EXTRACTION PROCESSING MANUFACT
LIGHT/POWER UTILITIES
• LUNETTA
• FUSION POWER
• MICROWAVE POWER SYST
• POWER RELAY SAT
• SOLETTA
FACTORY
12-HR ORB
ENERGY SYSTEMS
GEOSYNC ORBITS
SOLIDIFICATION PRODUCTS
• DC REGULATION CRYSTALS
• HI-TEMP TURBINE BLADES
• HI-QUAL TUNGSTEN CARBIDE
• SPECIAL METAL COMPOUNDS
• HI-STRENGTH MAGNETS
• NEW/BETTER GLASSES
• MICRO-ELECTRONIC DEVICES
• BUBBLE MEMORIES
• NEW/BETTER BEARINGS (E.G. Al-Pb)
• SOLAR CELLS
FACTORY
SUN SYNC ORBITS
BIOCHEMICAL PRODUCTS
• UROKINASE
• ISOENZYMES
• ANTIBIOTICS
• HORMONES
ETC.
• SPACE PHARMACOLOGY
• NEW HEALTH SERVICES
INFORMATION ACQUISITION
NEAR-EARTH ORBITS
LIFE SCIENCES
NEW ECOSPHERES
* HABITATS
* BIOSATELLITES
• BASIC LIFE PROCESSES
• MEDICAL STANDARDS
• ADV LIFE SUPPORT SYST
• MEDICAL STATIONS
• CONTROLLED PHYSIOLOGICAL & GENETIC ENGINEERING
• STANDARDS FOR LARGE HABITATS
AREOPHYSICS
LUNAR ORBIT
INFO TRANSMISS
ELECTRONIC SERVICES
• DOMESTIC SAT. SYSTEMS
• EDUCATION/INFO SATS
• MEDISATELLITES
• BUSINESS SATS
• MAIL SATELLITES
• MARITIME SATS
• COM TO SMALL MOBILE TERMINALS
• INTERLINK SATS
• INTER-WORLD COM NETWORK
EARTH, SUN COMPARATIVE PLANETOLOGY
• WEATHER
• LAND/AGRIC/FORESTRY
• FRESH WATER CYCLES
• OCEANS
• ATMOSPHERE
• CLIMATE
• GEOPHYSICS
• SOLAR PHYSICS
• AEROPHYSICS

Figure 1

(Figure 2). The possible or probable stabilization pattern of the four different population groups shown in Figure 3 adds up to global stabilization toward the end of the next century at a level around 11 billion. Precisely because this demographic factor is long term, a more rapidly acting agent is needed. This is technology and industrialization. But accelerated industrialization can be accomplished best, without unnecessarily severe damage to the terrestrial environment, by taking maximum advantage of space industrialization.

KEY SOCIAL BENEFIT POTENTIAL OF SPACE INDUSTRIALIZATION

- REDUCES THE POSSIBILITY OF RESOURCE DISTRIBUTION WARS (AVOIDANCE OF SHRINKING WATER DROPLET WORLD SCENARIO)
- CONTRIBUTES TO RAISING THE QUALITY OF LIFE ON GLOBAL SCALE (AVOIDANCE OF ADMINISTERING SPREADING SCARCITY THROUGH ENHANCED CAPACITY FOR CREATING WEALTH)
- PROMOTES GLOBAL POPULATION STABILIZATION (AVOIDANCE OF POPULATION CONTROL BY DEATH THROUGH POPULATION MANAGEMENT BY QUALITY OF LIFE IN NEW SOCIO-ECONOMIC ENVIRONMENT FAVORING SMALLER FAMILIES)

Figure 2

Exoindustrialization does not mean large colonies in space, because they make evolutionary (socio-economic) sense only after the potential of exoindustrialization is reached. First we will need to find out what we can do to use weightlessness and the vacuum of space. By developing observation and communication systems in space we can begin to think of phases that will justify economically the involvement of human beings in space in large numbers. They will be in space on a strictly industrial enterprise. For reasons of economic viability no more people will be employed in space than are necessary. For a long time jobs in space will be quite expensive. A person having such a job will need to be very productive, and that will be possible only with the aid of automated devices. To sit down

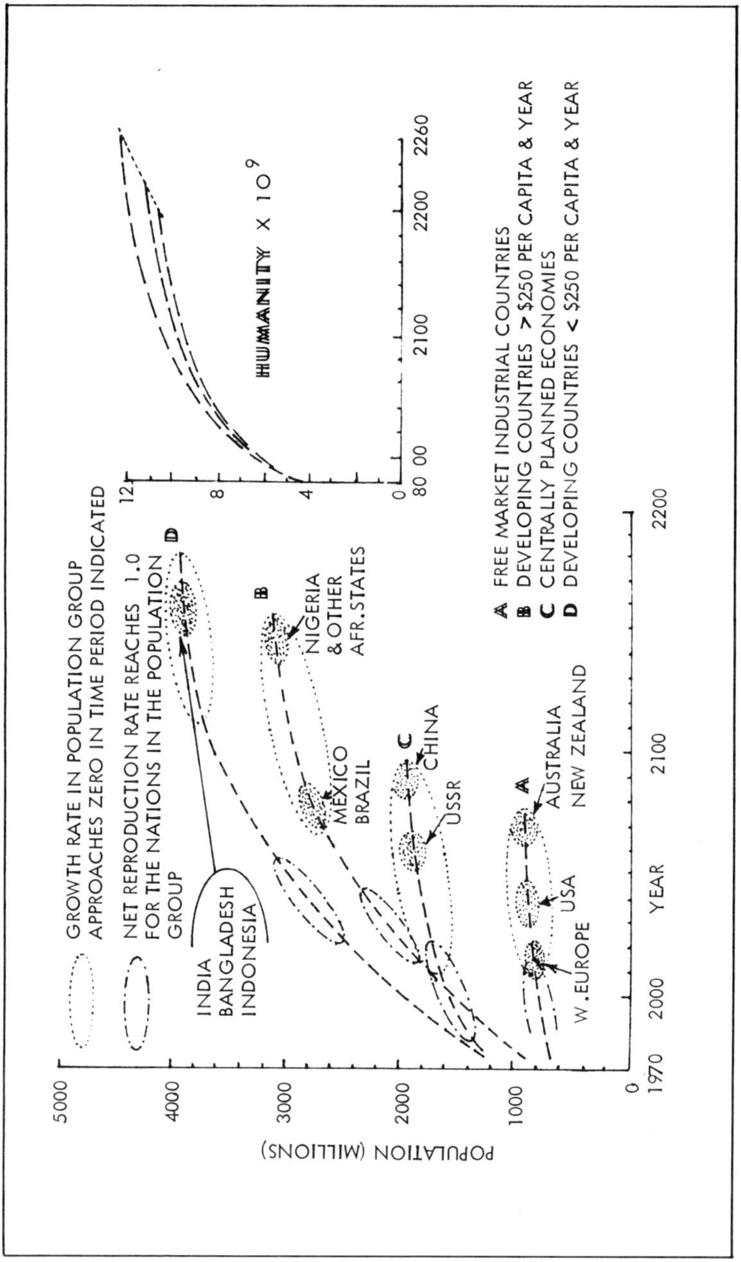

GROWTH RATE IN POPULATION GROUP APPROACHES ZERO IN TIME PERIOD INDICATED
NET REPRODUCTION RATE REACHES 1.0 FOR THE NATIONS IN THE POPULATION GROUP
INDIA
BANGLADESH
INDONESIA
D
B
NIGERIA
& OTHER
AFR. STATES
MEXICO
BRAZIL
C
CHINA
USSR
A
AUSTRALIA
NEW ZEALAND
USA
W. EUROPE
POPULATION (MILLIONS)
5000
4000
3000
2000
1000
0
1970
2000
2100
2200
YEAR
HUMANITY X 10^9
12
8
4
0
80
00
2100
2200
2260
A FREE MARKET INDUSTRIAL COUNTRIES
B DEVELOPING COUNTRIES > $250 PER CAPITA & YEAR
C CENTRALLY PLANNED ECONOMIES
D DEVELOPING COUNTRIES < $250 PER CAPITA & YEAR

Figure 3

and say we want a thousand people in space now and then have to find a job for them won't work. We have to do it the other way around.

Getting into space means we must learn to build in space, and to do that we need to be able to process materials and energy and to handle structures in space. As soon as possible we should be able to utilize lunar raw materials.

LUNAR INDUSTRY

In the geolunar industrial system the Moon plays a special role as the only Earth Space Station with indigenous raw materials. Its surface exceeds by about 10%, the area of Europe and Africa combined.

The Lunar Industrial Connection

A primary advantage of lunar industries is that they offer the option of separating production (elsewhere) from consumption (on Earth) in an industrial civilization where it becomes increasingly difficult to do both indefinitely in the same environment. This implies that the lunar environment could be used not only for material extraction but for refinement and manufacturing as well, within the context of economic viability.

A second major advantage of lunar industrialization is that it puts raw materials and industrial products (later probably also chemicals and biomaterials) close to orbital space, thereby opening yet another environment to production for terrestrial consumers and extraterrestrial markets. Large-scale space manufacturing can evolve only on the basis of lunar resources (to which certain terrestrial resources must be added). Orbiting factories near Earth are restricted to special products, because of the comparatively high energy requirement and cost of Earth to near-orbit delivery.

The Moon offers a low-g vacuum environment on the surface. Because of low gravity, the zero-g environment in circumlunar orbit is more easily accessible. Compared to Earth, it takes less than one-twentieth the amount of energy to transport a given mass from surface to orbit. Therefore, both lunar surface and the near circumlunar orbital space can be used advantageously as industrial processing space.

Seleneconomy

Lunar industrialization should be examined in the light of the requirements of the overall goals and objectives of geolunar industrial development with its terrestrial and extraterrestrial market potential for lunar products. These, in turn, determine the flow of energy and material through the lunar economic (seleneconomic) and geolunar industrial system.

The lunar industrial establishment probably will consist of two components — the surface installations of one or several Lunar Industrial Zones (LIZ); and an orbiting component or Lunar Industrial Space Installation (LISI). Of all exoindustrial activities in geolunar space, the lunar operations represent most closely a total industrial system. It is supported by carefully minimized import of non-indigenous materials from Earth. In turn, it supports the terrestrial and orbiting industry with purified materials and with high-quality zero-g and/or vacuum manufactured processed goods. Lunar industrial operations must demonstrate not only engineering and operational cost-effectiveness and virtually complete ecological autonomy but economic viability as well.

The gross lunar product of a stable lunar industrial economy must, through value generation (+$$ . . . .), be able to pay for needed terrestrial imports (-$), its own operating costs and generate a profit for financing expansions and advancements. The values generated are, to a smaller part, used for internal purposes. Mainly, it is assumed they could be sold to orbital markets and terrestrial consumers. Therefore, it is important that a lunar industrial capability be preceeded by the buildup of orbital markets — manufacturing facilities in near-Earth orbits, space stations and other space industrial facilities in geosynchronous orbits and intermediate orbits.

Lunar "Crude"

The characteristics of the lunar material resulting from the Apollo missions and the USSR unmanned probe missions have been studied in great detail. Based on sample findings so far, the lunar surface consists largely of at least three types of igneous (formed from a molten state) rocks. There is a mare basalt which is dark, iron-rich and relatively heavy. The anthrosite is lightly colored, aluminum-rich, less dense, and is the major type of rock found in the lunar highlands. The third type, the norite or gabbo is slightly denser than the anorthosite and contains mostly feldspar and other silicates.

The basic constituents of these lunar rock samples are the same as those found in the same kinds of Earth rocks and meteorites. The oxides of silicon, iron, aluminum, calcium, magnesium and titanium make up between 97 and 100 percent of the material. However, compared to terrestrial rocks, lunar rocks are enriched in some of the refractory elements, notably zirconium, titanium and chromium. The alkalis and most volatile elements are seriously depleted. The volatile elements on the Moon are only 1 to 0.3 percent as abundant on Earth, and there is just as little water (see Figure 4). Nitrogen and carbon are also found in the lunar fines (sand), however their abundance is very low: a thousand parts per million. For this reason, hydrogen or carbon will have to be imported from Earth — at least until lunar energy technology and processing capacity is very advanced.

The elements are not evenly distributed. This results in "metallogenic provinces," or regions in which certain metals are more abundant — for example, on a macroscale, the highlands are richer in aluminum, the maria (the dark lowlands) are richer in iron, and on a smaller scale, the Mare Tranquilitatis region in which Apollo 11 landed appears to be a titanium province. Representative abundances of the major oxides are shown in Figure 5 where each circle indicates the Apollo mission number and destination.

In spite of the fact that oxygen amounts to almost half of the mass of the lunar sample material (about 43 percent by

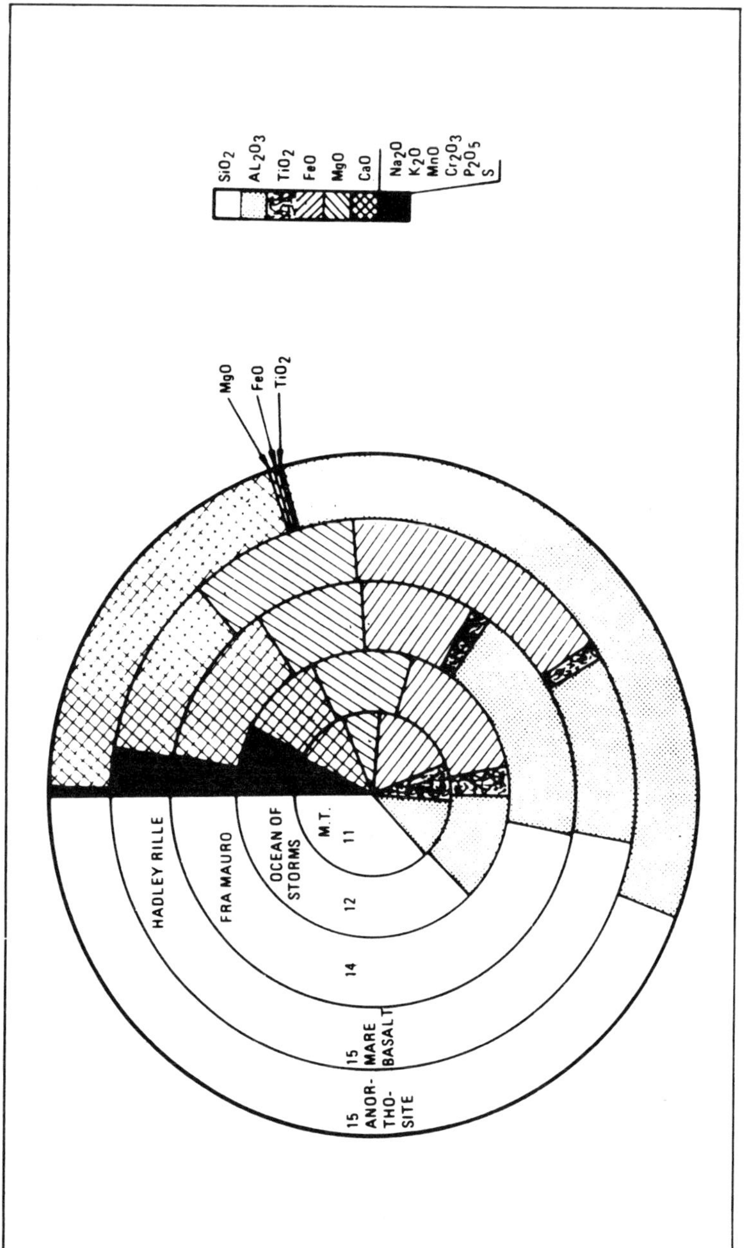
SiO2
AL2O3
TiO2
FeO
MgO
CaO
Na2O
K2O
MnO
Cr2O3
P2O5
S
MgO
FeO
TiO2
HADLEY RILLE
FRA MAURO
OCEAN OF STORMS
M.T.
11
12
14
15 MARE BASALT
15 ANOR- THO- SITE

Figure 4

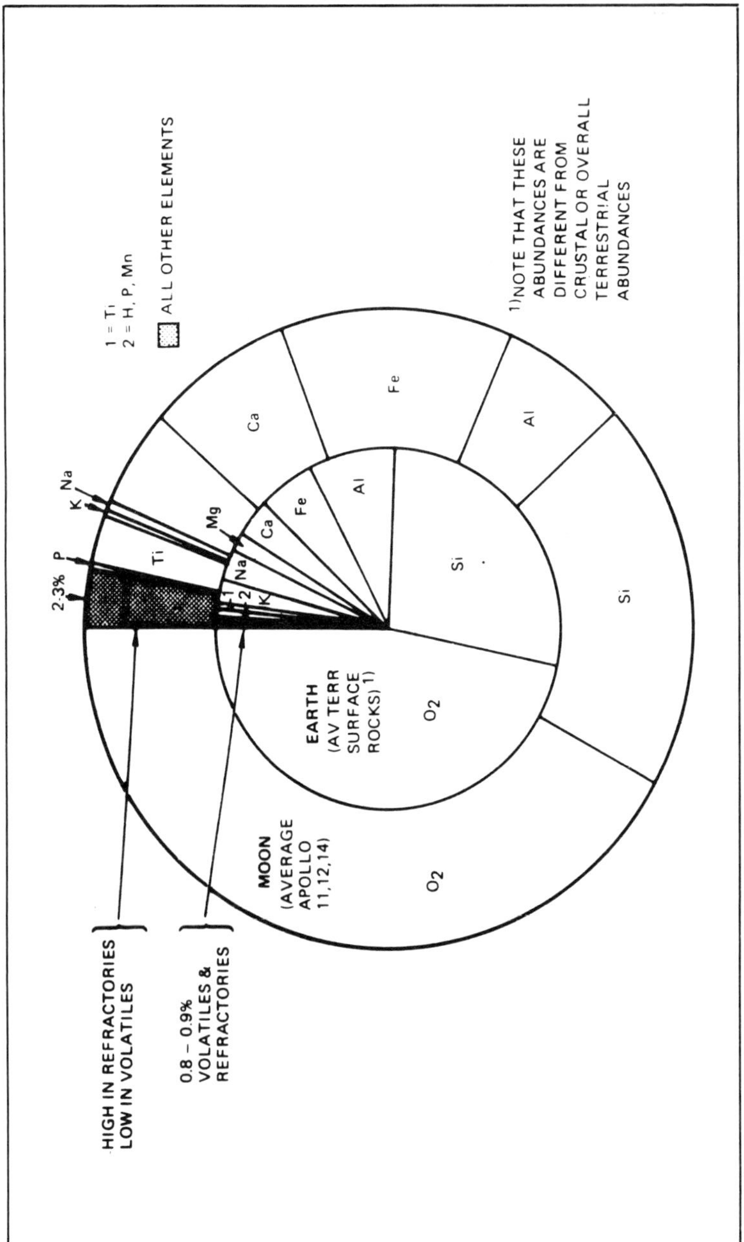

1 = Ti
2 = H, P, Mn
ALL OTHER ELEMENTS
1)NOTE THAT THESE ABUNDANCES ARE DIFFERENT FROM CRUSTAL OR OVERALL TERRESTRIAL ABUNDANCES
EARTH (AV TERR SURFACE ROCKS) 1)
MOON (AVERAGE APOLLO 11,12,14)
O2
Si
Al
Fe
Ca
Mg
Na
K
Ti
P
2-3%
HIGH IN REFRACTORIES LOW IN VOLATILES
0.8 – 0.9% VOLATILES & REFRACTORIES

Figure 5

weight), lunar materials show an apparent oxygen deficiency (i.e. they are not fully oxidized) of about 1.6 percent as compared to the type of saturation that can be assumed for terrestrial rocks.

Lunar Industrial Processes

If the full spectrum is achieved, lunar industrial processes begin with mining and end with assembly of manufactured parts. Figure 6 offers a synoptic overview of lunar industrial processing. On the right are shown transports arriving and leaving. In addition, deliveries to the lunar surface may be made from orbiting facilities. Moreover, "waste" from deliveries may be useful materials on the surface. In an attempt to be as efficient as possible, packaging materials, (including propellant tanks no longer used for their primary purpose) can be made to represent a valuable source of raw materials that are scarce on the Moon.

This means that new and different guidelines will be adopted for the material composition of packaging material. Plastics will be useful as biomaterials and possibly as a source of industrial hydrogen and carbon on the Moon. Wherever reasonably applicable, tin containers should be used rather than aluminum containers, since tin is an industrially useful metal and rare on the Moon, whereas aluminum is not. Zinc and copper should be added to alloys for other containers and fluid tanks. Zinc is vitally important for plants and animals and also has industrial applications. Wherever reasonable, carbon containing materials should be used as supporting structures and in packaging arrangements of payloads because of the fundamental biological and industrial importance, and lunar paucity, of carbon. The delivery of hydrogen and nitrogen may be combined in ammonia which can be transported at more convenient temperatures than either component. These are only a few examples of how one can re-think the material technology of packing and other supporting materials as well as of liquified gases.

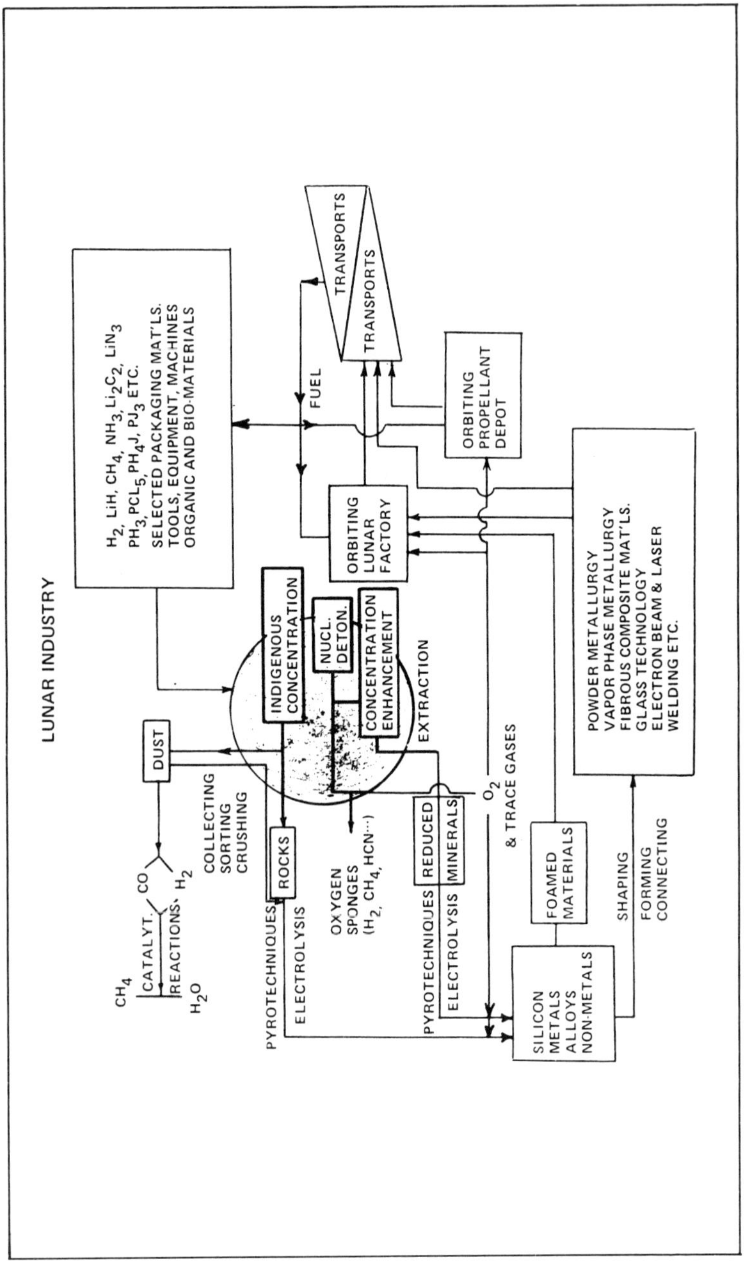
LUNAR INDUSTRY
H2, LiH, CH4, NH3, Li2C2, LiN3
PH3, PCL5, PH4J, PJ3 ETC.
SELECTED PACKAGING MAT'LS.
TOOLS, EQUIPMENT, MACHINES
ORGANIC AND BIO-MATERIALS
TRANSPORTS
TRANSPORTS
FUEL
ORBITING
PROPELLANT
DEPOT
ORBITING
LUNAR
FACTORY
INDIGENOUS
CONCENTRATION
NUCL.
DETON.
CONCENTRATION
ENHANCEMENT
EXTRACTION
DUST
COLLECTING
SORTING
CRUSHING
CH4
CATALYT.
REACTIONS
H2O
CO
H2
ROCKS
PYROTECHNIQUES
ELECTROLYSIS
OXYGEN
SPONGES
(H2, CH4, HCN···)
REDUCED
MINERALS
PYROTECHNIQUES
ELECTROLYSIS
O2
& TRACE GASES
FOAMED
MATERIALS
SILICON
METALS
ALLOYS
NON-METALS
SHAPING
FORMING
CONNECTING
POWDER METALLURGY
VAPOR PHASE METALLURGY
FIBROUS COMPOSITE MAT'LS.
GLASS TECHNOLOGY
ELECTRON BEAM & LASER
WELDING ETC.

Figure 6

Returning to Figure 6, we see two major methods of extracting materials from lunar crude. In the first method, based on indigenous concentrations, the lunar materials are processed in their original condition. The lunar dust is collected and heated to release hydrogen and other gases, while metals are extracted from the lunar sand and crushed rocks by electrolytic extraction.

To be economically competitive, industrial raw material extraction requires comparatively large scale production, productivity and low-cost energy. This means that equipment/facility cost, transportation costs and operating costs should be low.

A lot of equipment is needed for these methods to collect, sort, crush, if necessary, and process lunar surface material and to provide the needed electricity or heat, using solar energy. Figure 7 shows a solar powered mobile lunar industrial vehicle (LIV). Smaller LIV's may be used initially. They can be remotely piloted from the shirtsleeve environment of a habitat or even from Earth with temporary control overrides from circumlunar control facilities. Initially, when production requirements are not yet very large, the vehicle may be equipped with a solar-thermal Brayton cycle electric power generator, rendering it independent of a major central power system. The power is used alternately to drive the vehicle, to operate its collection and sorting mechanism, to generate heat, to power an electrolytic melt for metal extraction and to electrolytically decompose water into oxygen and hydrogen as a power source for operation during the lunar night. Substitution of the solar energy by nuclear energy in a second-generation LIV greatly increases its productivity because the long lunar night is a major obstacle to the productive use of solar energy.

A solar or preferably nuclear powered LIV, containing one or several electrolytic furnaces and centrifugal separators, can, as it moves along, scoop up lunar regolith (the layer of loose rock material), produce oxygen and separate the various metals contained in the regolith. Mobility and the loose regolith render it unnecessary for the vehicle to dig deep. It's essential-

Figure 7

ly a strip mining device.

For still higher-capacity mining operations and use of lunar crude, the LIV's are replaced by a semi-stationary system (Figure 8). Arrays of large electrolytic and centrifugal furnaces are set up, powered by a stationary nuclear power plant of multi-megawatt capacity. A strip mining area is identified. Collector trucks strip mine regolith and load the material on a conveyor belt consisting of modules that can be moved so that the system can be relocated to service other strip mining zones. The collector trucks are electrically powered. The electricity is provided by the nuclear plant through overhead lines from which the trucks draw power via flexible connectors. While solar power may be used for supplementary purposes during the lunar day, the nuclear power base permits around-the-clock operation with correspondingly high production.

The second method is concentration enhancement and is based on nuclear underground detonation. The thermal energy of the detonation separates the oxygen from the metals and silicon. Removal of most of the oxygen leaves enhanced metal concentrations behind which then can be mined. Nuclear explosives are of particular interest for large-scale production with no effect on the vacuum environment, if detonated underground. They provide the highest presently available concentration of energy per unit mass. They are also cost-effective without environmental dangers and biospheric conflicts. Nuclear fission (or preferably fusion) charges can be the key to economic raw material extraction, thereby reducing the burden on our Earth environment.

Lunar Products

In 1972 I presented in a paper before the International Astronautical Congress in Vienna an extensive list of lunar industrial products. With minor changes, this list is shown in Figure 9. On the outer ring are located products of mining, benefication and refining processes — such as oxygen, geoscarce metals (not necessarily in the absolute sense of the term, but in terms of economic/political availability to countries or regions

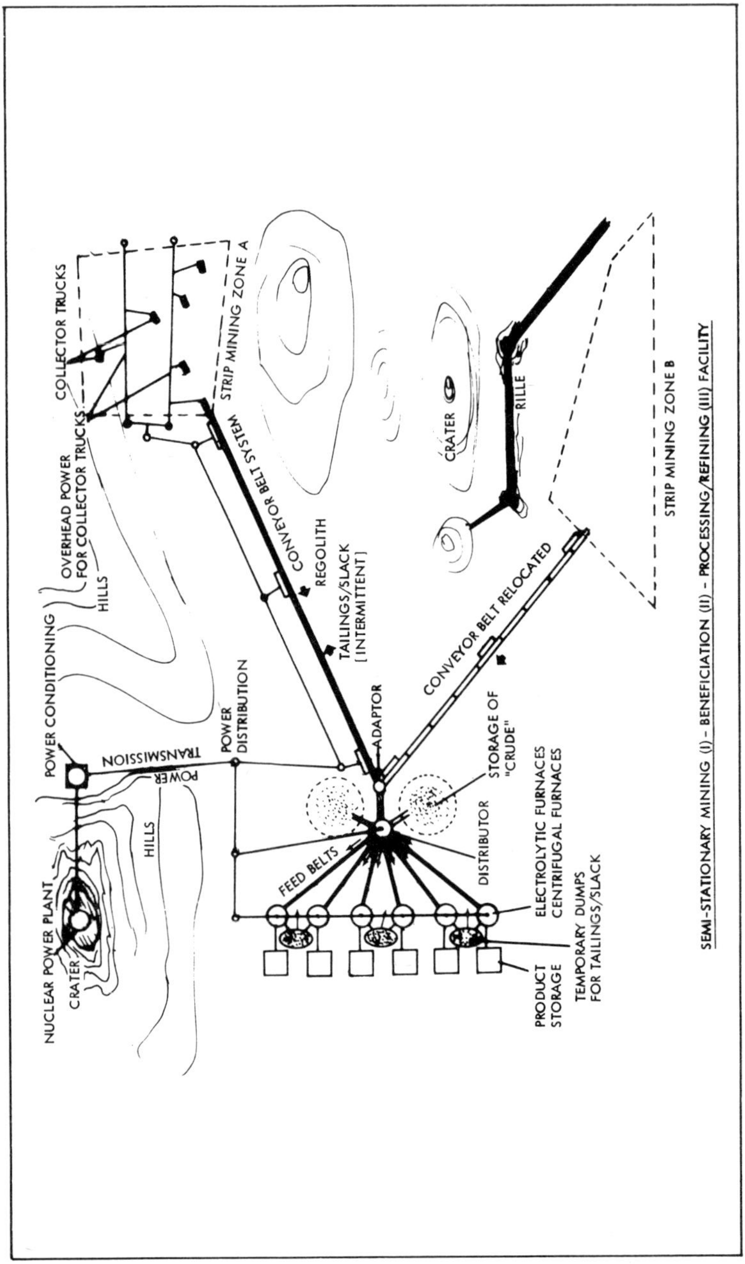
NUCLEAR POWER PLANT
CRATER
HILLS
POWER CONDITIONING
POWER TRANSMISSION
POWER DISTRIBUTION
HILLS
OVERHEAD POWER FOR COLLECTOR TRUCKS
COLLECTOR TRUCKS
STRIP MINING ZONE A
CONVEYOR BELT SYSTEM
REGOLITH
TAILINGS/SLACK [INTERMITTENT]
ADAPTOR
CONVEYOR BELT RELOCATED
STORAGE OF "CRUDE"
CRATER
RILLE
STRIP MINING ZONE B
FEED BELTS
DISTRIBUTOR
ELECTROLYTIC FURNACES
CENTRIFUGAL FURNACES
TEMPORARY DUMPS FOR TAILINGS/SLACK
PRODUCT STORAGE
SEMI-STATIONARY MINING (I) - BENEFICIATION (II) - PROCESSING/REFINING (III) FACILITY

Figure 8

lacking corresponding resources within their own territory), geo-abundant metals, silicon, silica glasses, rock and uranium (also thorium) as well as isotopes (either indigenous or deposited earlier as part of terrestrial nuclear waste, cf. below).

The inner ring refers to the subsequent lunar industrial utilization for manufacture of semi-finished products (sheet metal, castings, bars, wire, powders of pure or alloyed materials, using lunar vacuum and low gravity, or the micro-g conditions of manufacturing in CLO (Circumlunar Orbit), part of the overall lunar industrial sphere, crystal bole, glass wood and fibers), parts, components, sub-assemblies and full assemblies, carried out on the lunar surface or in CLO facilities.

Products of the outer and inner rings, and their markets are listed outside the rings. The markets include three major groups, and this is valid also today: transportation systems, orbiting facilities and terrestrial customers (industries, jewel trade and art trade).

Lunar Transportation

A discussion of concepts for low-cost transportation through geolunar space is involved and would exceed the frame of this presentation. Instead, we will address the subject of low-cost access to and from the lunar surface.

In a conventional landing, retro-rockets break the descent near the surface. Gases are released into the lunar environment. At heavy traffic, this may result in an undesirable accumulation of gases, especially the chemically aggressive oxygen. Moreover, for reasons of cost, it is important to consume as little propellant as possible during the descent and ascent. The slide landing is a novel landing concept particularly attractive for two reasons: (1) a slide landing can almost eliminate the lunar gravity potential as a source of propellant expenditure supplied from Earth; and (2) it can preserve the high vacuum conditions at the lunar surface. The vehicle, so to say, lands and slows down on the surface of the lunar glass ocean. Figure 10 shows the frontal view of the lunar slide lander concept. The right demonstrates the final slide phase at still high speed

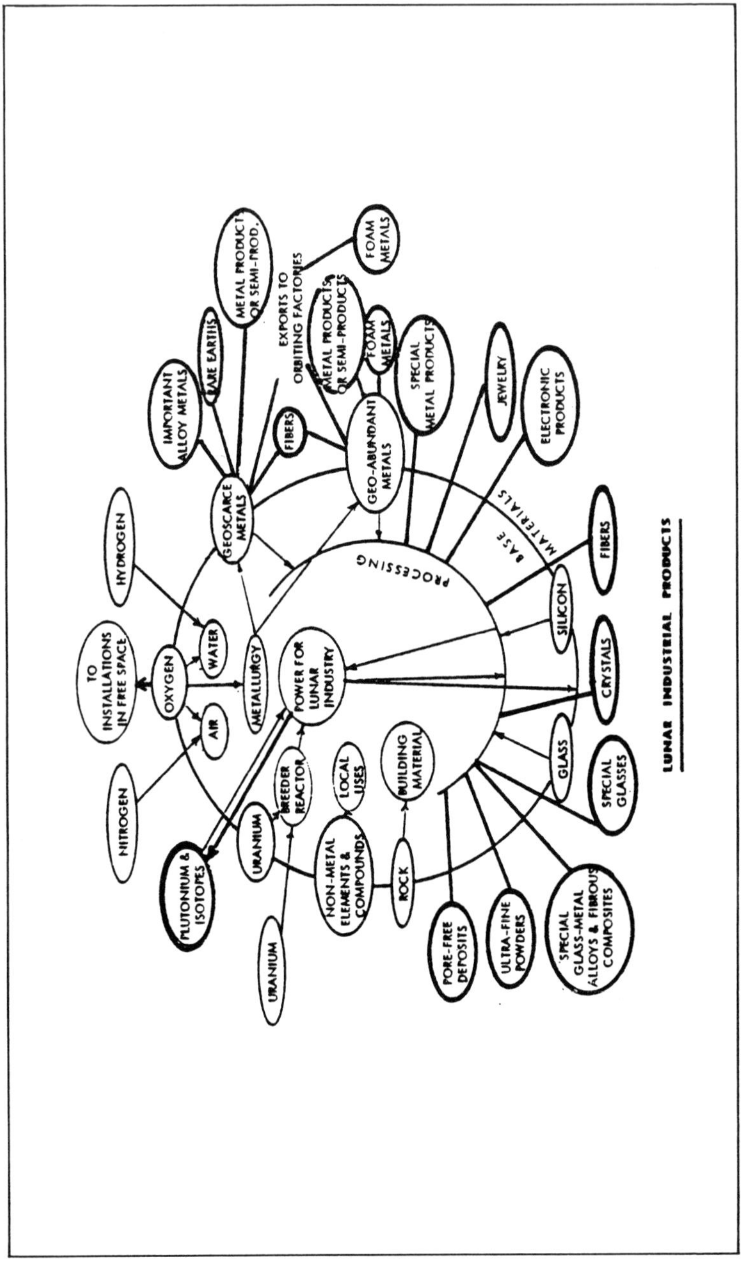
LUNAR INDUSTRIAL PRODUCTS
PROCESSING
BASE MATERIALS
NITROGEN
HYDROGEN
TO INSTALLATIONS IN FREE SPACE
OXYGEN
AIR
WATER
METALLURGY
POWER FOR LUNAR INDUSTRY
GEOSCARCE METALS
IMPORTANT ALLOY METALS
RARE EARTHS
METAL PRODUCTS OR SEMI-PROD.
EXPORTS TO ORBITING FACTORIES
FOAM METALS
METAL PRODUCTS OR SEMI-PRODUCTS
FOAM METALS
FIBERS
GEO-ABUNDANT METALS
SPECIAL METAL PRODUCTS
JEWELRY
ELECTRONIC PRODUCTS
FIBERS
SILICON
CRYSTALS
GLASS
SPECIAL GLASSES
SPECIAL GLASS-METAL ALLOYS & FIBROUS COMPOSITES
ULTRA-FINE POWDERS
PORE-FREE DEPOSITS
ROCK
BUILDING MATERIAL
LOCAL USES
NON-METAL ELEMENTS & COMPOUNDS
BREEDER REACTOR
URANIUM
URANIUM
PLUTONIUM & ISOTOPES

Figure 9

Figure 10

hurling streamers of glowing lunar fines to both sides.

Slide landing requires high, but attainable navigational accuracy, a complex sensor system, rapid control response and it must be computer controlled in view of a touch down speed of about 6000 km/h. The lander can be piloted or remotely controlled from a flight control facility.

The slide lander is kept on its required path by a guide beam system along the slide path. The slide path is a specially prepared landing strip. The lunar surface conditions are inherently favorable to the slide landing mode. The ground is covered with a layer of rubble (regolith). A relatively soft surface is therefore provided naturally and it remains primarily to control the particle size to a depth of 1 to 3 m by removing stones above a given size. The maria are generally smoother and less heavily cratered than the highlands. The choice of location must take into account the shallow approach path and must also avoid higher mountains downrange, in order to accommodate re-ascent if the slide landing maneuver is not negotiated.

The adhesion of lunar dust particles, due to vacuum conditions, is not a disadvantage. In fact, it appears advantageous inasmuch as it gives the dust a more water-resembling cohesiveness. The momentum of the lander is too large to be deflected by cakiness of lunar dust, or by small rocks.

With a length of 80 to 40 km and a width of 100 m, the landing strip is of enormous proportions by terrestrial standards. But, given the nature of the surface and its basic simplicity, the construction job is simple and predictable and requires not much more than the development of the sweeper, which does not involve particularly advanced technology.

The slide landing mode promises a significant further reduction in Earth supply requirements, hence, increased self-sufficiency. While the construction job is basically simple, the landing technique contains advanced elements.

Through slide landing, the lunar surface can become as accessible as a station in lunar orbit. It will pave, in the post-initial phase, the way to the development of a lunar industrial economy that will support the first stage exosettlements.

Ascent from the lunar surface to CLO (Circumlunar Orbit)

will be accomplished initially by the conventional mode. But, in the present conceptualization of lunar industrial development, the ascent transportation system will be called upon to move a variety of payloads — metals and other solids, liquid oxygen and personnel. Besides the need to preserve the lunar high vacuum, it is a waste to dissipate as valuable a combustion product as water (assuming O_2/H_2 propulsion). At first, this exhaust material can serve a second purpose: the accumulation of water for the lunar habitats and industrial purposes. This is a more economical use of the hydrogen carried to the surface than to dissipate it. Later, the captured exhaust water can be decomposed electrolytically to oxygen and hydrogen and recycled as propellant many times over, at corresponding savings in transportation costs for not having to import more than very small amounts of make-up hydrogen, considering some unavoidable losses. To achieve the goals for the use of hydrogen is the use of the Enclosed Launch Track (ELT).

The ELT contains the combustion gases, prevents their release into the environment and permits hydrogen recovery. If the O_2 were heated by laser energy beamed along the track, no electrolytic facility would be needed, only an electric power plant. The ELT complements the slide landing mode, since the slide lander is designed for horizontal take off. During the acceleration phase, the vehicle is again lifted off the ground by its vertical thrust modules.

As pointed out earlier, because of the uneven distribution of the elemental abundances on the lunar surface, it is unlikely that one single location will be ideal for processing. Moreover, the central processing facility (CPF) is also influenced by accessibility and by the topographic feasibility of slide landing. Moreover, it would not be cost-effective to build many central processing facilities. The solution to this problem lies in the establishment of "feeder stations," and ballistic delivery. If richer findings of certain materials exist at not too great distance from the CPF (up to several hundred km) single feeder stations can be established at those places and the raw material be delivered ballistically into a predetermined reception area (crater) close to the CPF. The CPF/feeder station system then

Figure 11

forms a lunar industrial zone (LIZ). Figure 11 shows a 300 km LIZ around a Tranquility facility, including, at least, titanium and aluminum rich areas. The map also indicates another zone of 300 and 600 km radius respectively. The minimum energy ballistic delivery profiles are indicated at the western and eastern limits. The launch velocities for hurling the lunar crude from feeder station to the CPF reception area are comparatively small — e.g. 671 meters per second for the 300 km distance. The feeder stations should be as simple and self-sufficient as possible. These two factors (simplicity and modest performance requirements) make the early introduction of electromagnetic launchers at the feeder stations particularly suitable (Figure 12).

Figure 12

Habitats on the Moon

For the construction of habitats, especially at the beginning, it is logical to take maximum advantage of what is there already, such as protective mountain sides, rills, craters, as well as lunar material suitable for construction. On Earth, ten tons of atmosphere per square meter, shield us effectively from solar radiation. A large portion of the terrestrial atmosphere consists of nitrogen, and nitrogen absorbs the slow neutrons that are created by the effect of cosmic radiation. On the Moon, in absence of nitrogen, shielding can be assured by using the protection that topography and lunar materials can furnish.

Habitats will start out small, but, in response to lunar economic growth will eventually expand into macrosettlements. I have studied one prototype named Selenopolis. Selenopolis consists of a number of huge covered semi-tunnels. Different sections have different climates. For example, one section could have a northern climatic zone with wheat, and birch trees (Figure 13). Other sections have Mediterranean and desert-type (dry, warm) climates. One can readily commute back and forth between these environmental modules. Where climatically appropriate, the modules go through seasonal changes to accommodate the plant ecology. The Russians have

Figure 13

grown various plants in lunar soil — wheat, barley, and cabbage, so we may not have to bring up Earth soil. But it does take time to get the lunar soil nitrogenized. The habitat will be illuminated by reflector systems using the Sun, and at lunar night by diffuse electric lights from nuclear fission or fusion power plants.

A major advantage of the lunar habitat system approach is that it permits a gradual build-up from small habitats for tens to a few hundred inhabitants to macro-habitats that eventually can house millions but do not have to be planned and committed at the initial experience level. People can develop their own way of life and build larger habitats based on their experience. If the people come to prefer in greater numbers a particular environment, climate or arrangement, then that particular system can be extended. According to economic growth as a necessary supporting base, the influx of lunar immigrants will vary. The habitat extension will accelerate or slow down accordingly. There is therefore no need to maintain more than is necessary. Each extension draws on preceding experience and is financed by earnings rather than begging terrestrial government for funds.

The Moon will be an environment where the Human is first. Creation of a biosphere will *begin* with us. On Earth, the biosphere was first. It gave birth to the Human species and human terrestrial activities must make allowances for this fact. On the Moon, the story will be different.

Living at one sixth gravity will be very pleasant. Astronaut Dave Scott praised the lightness on the Moon. He was also enticed by the lunar landscape which I, too, find very beautiful. On the Moon, one can go out for a walk. Space suits can be made much more comfortable and more convenient than those the Apollo astronauts have worn. I have seen advanced designs of space suits at Houston that would make walking on the Moon fun. Thus, on the Moon, there is a world outside in addition to the world inside for the habitants.

The lunar macro-settlements are "cells" of life in a new world with which they are integrated in many ways. They do not become mere make-believe Earth imitations, but new, characteristic entities of their own.

If a man of the last century preferred the convenience of the Court of Versailles or King James, we wouldn't find him on the California frontier. The man of the California frontier didn't go there to build another London or Paris. No, he went there because he was willing to embrace the new world as it was, and for its own sake. In fact, he was attracted by the difference. The wilderness, the unlimited free land, and the new life style became his new "home." Those who tried to transplant the arrangements and ways of the old world into the new, were unsuccessful and vanished, or at least became peripheral to the new history. Space habitats built to simulate Earth on a micro-scale conceptually miss the whole point of expanding living beyond Earth. If you go into space, it is because you are ready to accept something different. If you want Earth, there is nothing that can compare with Earth. When you choose to live on the Moon it is because the lunar environments appeal to you.

Lunar Development

The development of the Moon will have a humble beginning; but, as lunar productive capability takes hold, the developmental process will blossom into a grand orchestration of virtually all branches of human activity, applied for the first time to build a truly new world — the selenosphere, consisting of its three basic components: technosphere, biosphere and sociosphere (Figure 14). The development of the selenosphere will progressively expand the scope of the androsphere beyond the geosphere — extending nominally to the geosynchronous orbit — to include two worlds. It is now time to understand what the selenospheric development means in an evolutionary context. The Earth's biosphere is integrated, and controlled by the chlorophyll molecule; it is modularized. What happens in one place eventually drifts through our air currents to other places, and that is vital for the biosphere to stay alive. If it did not have this interconnection, this communicating characteristic, it couldn't exist. The androsphere is

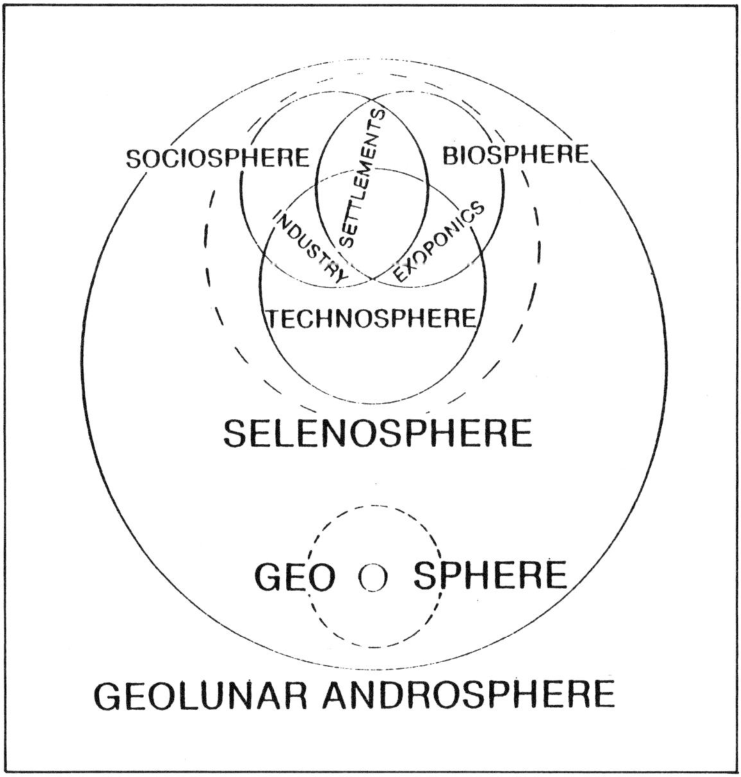

Figure 14

modularized. What happens on the Moon does not affect the Earth. Integration in the androsphere is done by the human mind. *We* are the new metabolic class and our energy gate is not the chlorophyll molecule but the human brain. With Earth, space, and other celestial bodies, a *physically open* world is established. It is integrated by this new metabolic life form, and as the biosphere is integrated by the chlorophyll molecule so this part of the open world, operationally and productively integrated, becomes the equivalent of the biosphere — an androsphere.

The biospheric characteristics are suited just fine for the photosynthetic activity. They are not suited so well for information metabolic industrial processes. The Earth is disturbed by our information metabolic industrial activity. It wasn't meant for its wastes and resource uses. In the androsphere we will be able to drive technologically in different environments and have control over what we want to introduce into the biosphere. A modularized open world is superbly suited for information metabolic industrial processes. In it, these industrial processes can unfold and reach levels of capability and productivity that could not be reached on Earth without seriously threatening or destroying the biospheric system. With extraterrestrialization, we witness a momentous evolutionary thrust of life as such into new growth potentials.

Within an airless extraterrestrial (second-order) subunit of the androsphere, such as the selenosphere, it is possible also to isolate the three basic components (techno-, bio-, sociosphere) from each other — at least for a much greater extent than in a subunit with an atmosphere and possibly also open water (i.e. not only Earth but also Mars and Titan, for example). In other words, while they are *functionally* integrated by the human brain, they can be *physically* isolated since there is an absence of a transporting medium like the Earth's atmosphere. However, in the third-order spheres — particularly the local biospheres and sociospheres — modularization will not necessarily be maintained. The creation of a socio-biological environment in a larger habitat is an example of non-modularized integration.

ANTHROPOLOGICAL DIVERGENCE

If you live most of your life on the Moon, or if you are born there, you will have a different bone structure because of the 1/6th gravity. The body is a chemo-physical system in equillibrium with its environment. This means that the physical condition, composition, tissue and bone structure are going to change somewhat to adjust to 1/6th gravity. Therefore, it is fortunate that Earth-gravity is higher than Moon gravity, because it is easier to go from a heavier gravity to a lighter gravity, than the other way around. Just imagine the Apollo astronauts landing somewhere where their body weight would be the equivalent of two thousand pounds!

In the late '50's, a film record was made of an Air Force experiment showing that chickens were raised ingeniously in a centrifuge, being fed water and food to grow up at different gravity levels. Some were raised at 2 g, others at 3 g. You should have seen those 3-g chickens! They were husky, with stong, stubby legs. The 2-g chickens were less bony, and the one-g chickens raised normally appeared almost gossamer compared to the 3-g chickens. These tests demonstrated a strong correlation between bone structure and g-level even in one generation. A generation or more of people raised on the Moon would have thinner, less calcium-containing bones and generally would develop a somewhat altered appearance.

In addition to the physiological divergence, there will also be an immunological divergence. The environment on the Moon will be somewhat different biologically from the environment on Earth. Earth is wildly alive with organisms. New bacterial or viral strains appear frequently. An extraterrestrial biosphere has to be more controlled. People living on the Moon might eventually view Earth as we today view ancient Rome or medieval Madrid. Were we transferred back to such environments, we would probably have diarrhea on the second day and in a week we would be dead. The dirt and the filth in which people of those times lived is unbelievable to us. Our immunity level today is different. We have to cope with less of some of the influences they were exposed to then, but we have to cope with others. Without any question, there is an indisputable difference in immunity. Lunar inhabitants, again, will establish different immunity levels and patterns.

There will also be aesthetic and cultural divergences, based on different appearances, different environments and differences in values and lifestyles. All these divergences, physical, immunological, aesthetic, cultural — will add up to a gradually broadening degree of anthropological divergence from terrestrial humanity.

The Moon is a base for a new extraterrestrial beginning. It provides the materials. It offers lower gravity. It has a great variety of different topographies which can be used to provide a high degree of protection against particle radiation. All this facilitates making a start and growing. The second and following generations, those who will become extraterrestrial, will build themselves a new world at the shores of space, preparing them to expand further into the solar system. With a fusion energy source controlled by information metabolism from the human brain, and aided by industrial metabolism, they will be able to move throughout the solar system. Photosynthetic life will operate where there is light, sunlight or light from other man-made sources. Materials will be found everywhere. So these people might build themselves a planetella, a small planet that may be viewed as a "cell" of humanity — an androcell.

ANDROCELL

An Androcell is an orbital, navigatable macro habitat with large horizontal cylinders parallel to the axis of revolution. (Figures 15, 16). In the main traffic and business arteries — they may measure as much as a kilometer across. Higher gravity regions (0.4 to 0.6 g) are located farther out for habitation; lower for agricultural, medical, and other purposes. Much of what is learned on the Moon will be incorporated into the system of the Androcell. As the Androcell voyages through the solar system, it will be powered by fusion energy — its own sun so to speak — that will be used as well for all other energy needs of the community.

Apartments will be like highrises with several recreational systems. There may be flight at zero gravity. There can be rotating lakes where the waters run uphill so that you can swim any way you want, and there may even be three dimensional tennis games.

The Androcell as a whole may be regarded as as new macrolife form, capable of existing in space indefinitely with its own protective and supportive arrangements. However, it is not an isolated island. It maintains mining bases and habitats on other worlds. It interacts information metabolically with other

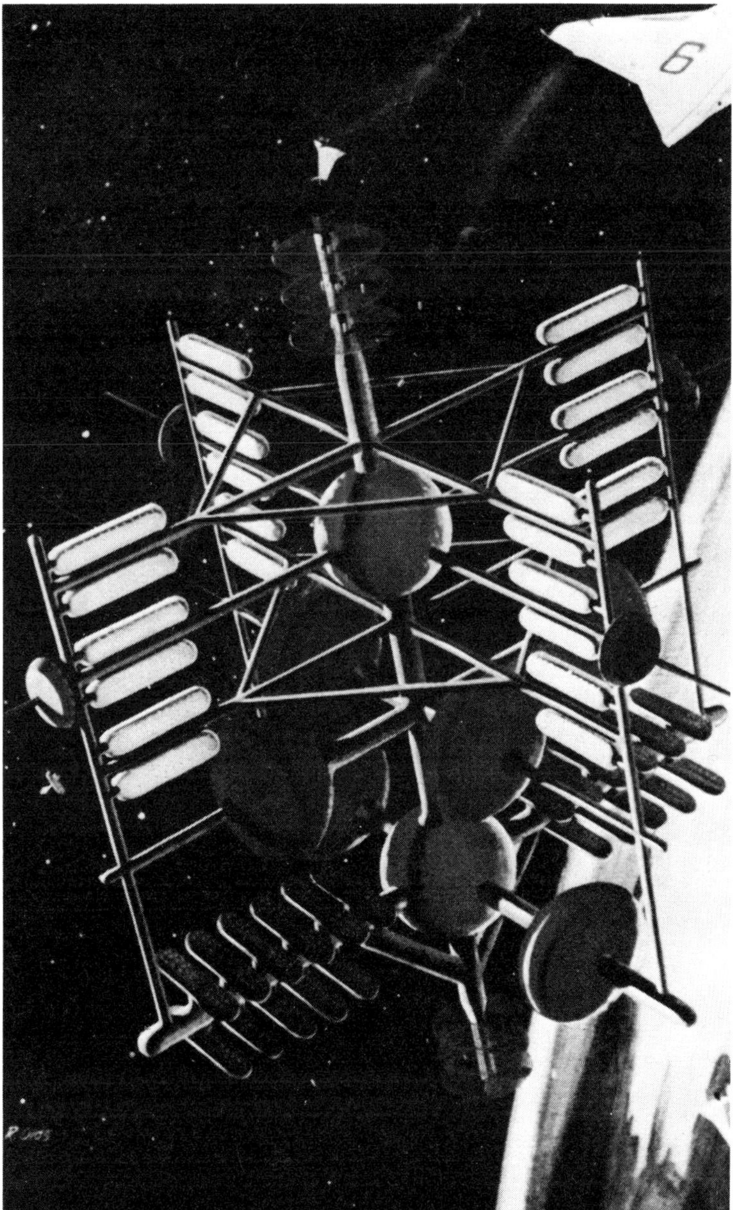

Figure 15

Figure 16

worlds through the human brain and industrially with other worlds through its fusion energy source — the sun it carries. In a sense, it is a floating solar system within the solar system. Its people will have a life of exploring other worlds, experimenting on their surfaces, establishing settlements, cultures and industrial bases. However, it is no longer part of space industrialization by Earth for Earth. It is a step beyond — the new life style of extraterrestrialized communities. It represents a new degree of anthropological freedom and plurality, as a result of finally being able to interact on a comprehensive scale with primordial matter for industrial and biological needs through the gate of human Information Metabolism.

Let us understand what is happening here in an evolutionary context. It is not unrealistic that in a generation (i.e. within 60 years or so) there will be interstellar flight, using very large Androcells. If so, these communities will experience the ultimate form of extraterrestrial civilization. With growth beyond the solar system, these societies psychologically outgrow dependence on a star-planet system. They must master anthropologically the ultimate desert — interstellar space — in which they will have to live for many generations. I don't know exactly how they are going to accomplish this, because their mentality will already have been changed so much through lunar and interplanetary experiences. But I am certain of one prerequisite — they must first learn to live (and I mean live, not just visit) in the outer solar system. The outer solar system simulates the interstellar environment physically and psychologically. It is a vast technological/anthropological testing gound for the ultimate journeys, where interstellar-type Androcells can be tested and their communities be adapted to the technology and psychology of complete energy, material and societal autonomy between stellar "landfalls." Yes, the outer solar system will "extrastellavite" those who press on into the universe, just as in an earlier stage the Moon will have been the testing gound and the cradle of extraterrestrialization.

There is no limit to growth as long as this universe is what it is. What we understand makes the future worthwhile, indeed, beyond our limited comprehension. It is up to us to assume this future, not to deny it to posterity by polluting it with cultural pessimism, no-growth nonsense, or Earth-parochial approaches and legalities. It is time to be born into a new open world — to accept once again the *extraterrestrial imperative*.